CONVERGENCE

CONVERGENCE

OUR WORLD IS CHANGING RAPIDLY
WHAT CHOICES WILL YOU MAKE IN THE DAYS TO COME?

LEWIS E. HILDRETH

© 2013 by Lewis E. Hildreth. All rights reserved.
2nd Printing 2014

Trusted Books is an imprint of Deep River Books. The views expressed or implied in this work are those of the author. To learn more about Deep River Books, go online to www.DeepRiverBooks.com.

No part of this publication may be reproduced, stored in a retrieval system, or transmitted in any way by any means—electronic, mechanical, photocopy, recording, or otherwise—without the prior permission of the copyright holder, except as provided by USA copyright law.

The author of this book has waived a portion of the publisher's recommended professional editing services. As such, any related errors found in this finished product are not the responsibility of the publisher.

Unless otherwise noted, all Scriptures are taken from the *Holy Bible, New International Version®, NIV®*. Copyright © 1973, 1978, 1984 by Biblica, Inc.™ Used by permission of Zondervan. All rights reserved worldwide. www.zondervan.com

ISBN 13: 978-1-63269-133-0
Library of Congress Catalog Card Number: 2012912015

CONTENTS

Preface vii

1. Transhumanism 1
2. Cyborgs 7
3. Genetic Engineering 17
4. Synthetic Life 29
5. Information Technology 39
6. Nanotechnology 45
7. Terrorism, Political Chaos, War 51
8. Weapons of Mass Extinction 63
9. The Population Time Bomb 75
10. Our Fragile Planet 81
11. The Angry Earth 91
12. It's the Economy, Stupid 103

13. UFOs to the Rescue! 113
14. Who Are They? 127
15. The Prophecies Reveal ... It *Is*
    Going by the Book 135
16. The Conclusion........................... 147

PREFACE

I ONCE VIEWED a documentary concerning the building of the first nuclear weapon. It was a large, spherical, metal encasement. What the theory scientists hoped would work was to place high explosives all around the perimeter of the sphere to be detonated at the same time. The belief was that if enough explosive pressure was exerted, or caused to *converge,* toward the center of the sphere where the uranium 235 was located, it would set off a chain reaction within the uranium—causing nuclear fission. It worked, and the nuclear weapon was born.

In like manner, I see a multitude of things *converging* from many different directions that will set off a reaction ... having unprecedented social, spiritual, political, and economic implications for our future and the future of the planet.

One day I read an article on how different schools of sciences were "converging" their technologies. By studying

VIII · CONVERGENCE

sometimes overlapping fields of scientific study, those who specialized in each field of the sciences were enhancing and advancing their own branch of scientific endeavor. What they soon realized was that the whole, the end product, was greater than the sum of its individual parts.

The different things I have observed and studied in the past decade had all been singular, unrelated events in my mind until I realized that there are also other issues running along in parallel lines, pointing to identical outcomes ... all concerning humanity's future and fate.

I concluded that these events are all related to each other, even though they are extremely diverse and *seemingly* unrelated. I realized that the sum of each of their individual parts was less than the total output they would produce when combined. The total output was subliminally pointing to an outcome far greater than what each individual field of research was pointing toward. The information that all of these "converging lines" were broadcasting had been staring me in the face all along; I had simply not "connected the dots."

The word that finally put it all together for me was *convergence*. I compiled a list of some forty topics I feel are converging to bring about the demise of planet Earth and the human population that resides on it. Unless God intervenes (and I believe He will), *humans will either cease to exist or be unrecognizable as a God-created species.*

A popular maxim says "sensationalism sells!" What kind of words or phrases could I use to "sensationally" describe what is going on in our world at this "critical hour" of history? Words and phrases like "time bomb," "dangerous," "explosive," "staggering," "terror," "frightening," "clueless," "unaware," and "terrible times" went through my mind. All of these words are designed to evoke emotion and draw the reader into wanting to read more, because it sounds exciting ... in a detached sort of way.

The truth is, *all* of these "sensational" words describe the current condition—or soon-to-be current condition—of what is going on in the world today. The world is a "time bomb" ready to explode. The people of the United States, as well as most other people of the world, do not have any idea about the convergence of the hidden agendas that dominate the laboratories, along with the political and economic "think tanks" located in most developed countries. The masses, for the most part, have been asleep and are totally ignorant of what they will soon confront; they are totally unprepared for what is soon coming.

In the following chapters Bible quotes will be used sparingly. Yes, there is the risk of having many readers reject this use, perhaps close the book in disgust or anger and not read any further. If this is you, here is a fair challenge: please consider all the prophetic sources you have relied upon until now and at least allow yourself to consider the one source you may have previously rejected (or perhaps never considered), *before* you make a judgment about it. Who knows? You might be surprised by the congruency of prophecy in Scripture with world events. Whether you accept or reject the Bible as authoritative, whether you accept or reject what it says is going to happen in the future, does not change any of the converging issues that are threatening the survival of our planet. These converging issues keep marching on, regardless.

X • CONVERGENCE

So, what is coming upon the Earth? What on earth is going on? Following is a sample list of converging issues—some are scientific, some biological, some technological, some natural disasters, some biblical prophecies, some social, some moral, some political, some spiritual, and even some "extraterrestrial."

For example, there are four technologies that are assembling their combined knowledge in order to enhance each individual field of endeavor and together create a new branch of scientific achievement. This achievement has, heretofore, only been in the imaginations of science fiction writers. These four technologies are genetics, robotics, artificial intelligence, and nanotechnology.

Related to these technologies is a school of thought and research that is known as *Transhumanism,* or posthumanism. Cyborgs, short for "cybernetic organisms," are one of the end products predicted. Included in this category is the creation of artificial or synthetic life. These and other scientific achievements and soon-to-be achieved experimentations will be discussed in greater detail in the following chapters.

As an example of what the scientific community has already achieved, there was recently a news release out of China. This article stated that Chinese scientists had spliced human genes into bovine embryos with the goal—and *achievement*—of having a cow produce human milk. The article went on to say that this product would be available in grocery stores in the next two to three years. For all intents and purposes, the cow looked like a cow.

The question is: "What did the failed experimentations look like?" After all, the cow *was* part human. The converse can also be said: the human was partly, if not mostly, cow. This then raises the questions, "What does it mean to be human?" and "What does it mean to be animal?" Obviously, it then follows that this raises not only religious, but moral and ethical questions as well.

Without getting too far ahead concerning the information written in the next chapters, this is a good place to note the following facts. *Every living thing has DNA*: humans, animals, plants, reptiles, sea creatures, insects, bacteria, viruses. DNA is the mechanism that stores all of the genetic information that makes each living creature unique. Since at least the early 1990s, scientists have been splicing the genes of one living creature into the embryos of a different living creature (as just explained with the cow giving human milk).

There is no limit to what can happen without moral, ethical, and religious "brakes" being applied. However, any of these "brakes" will only slow down the progress, not stop it. All living things can—and will—have their DNA exchanged to form some new unheard-of creature or humanoid.

Before going on with the sample list of convergences, I will quote Genesis 11:5–6 and comment on it. "But the LORD came down to see the city and the tower that the men were building. The LORD said, 'If as one people speaking the same language they have begun to do this, then nothing they plan to do will be impossible for them.'"

XII • CONVERGENCE

The LORD God said that if all the people speak one language, a language they could all understand, then *nothing* they plan (*nothing* they could imagine), would be impossible for them to ultimately achieve. God has told us what our capabilities are if we ever have or achieve one language. They are limitless!

The fact is, for the last 40 years, that "one language" *has* been developed! That language is now in full use worldwide. How can this be, you ask, since there are still many different, diverse languages being spoken around the world? The one common language is not linguistic, but *digital*. It operates on Base 2 math. The most common math today is Base 10, where everything is used in multiples of one through ten. Base 2 math operates on zeros and ones (0's and 1's). *It is the language of the computer.*

Computers speak to each other all over the world in this binary language. For example, if I would send an e-mail message to Japan, my computer (with the right software) would automatically convert all of my typed English into typed Japanese. This common computer language went worldwide in the early 1980s and it makes our language the same all over the world. Worldwide communication is instantaneous ... *and God said there is nothing we plan that will be impossible for us!*

Humans went from horse and buggy to landing a man on the moon, all within a 75-year period. What will the next 75 years bring? I believe there is urgency for people to be informed as much as possible at this critical hour in our history, thus the reason for this writing.

Getting away from the converging scientific topics and continuing with the sample list of convergences, the following examples are also pertinent to this discussion: Worldwide economic turmoil, worldwide social turmoil, energy shortages, population explosion, food shortages, a hatred of Israel by all nations, nuclear weapons proliferation, loss of morality, violence, a rejection of biblical authority by the Church, breakdown of the family, biblical prophecies being fulfilled—the list goes on. These topics will all be discussed in the coming chapters.

How much time remains? *I do not know.* Scientists give a timeline of 35 to 50 years to when "designer" humans, humans genetically-enhanced and/or interfaced with computer-enhanced physical and mental capabilities, will begin to be commonplace. Considering biblical prophecies and other converging topics listed above, *the timeline for history as we know it to end may occur much sooner.*

While many of the scientific technologies discussed in the first few chapters hold a variety of promising developments for the health and well-being of people, the fact remains that the uncontrolled use of those technologies *will* bring out the dark side of human nature. These all point toward the approaching end of this age. Many of these technologies have military application. The nations of the world cannot afford to fall behind in these technologies and so will pour billions of dollars into further research. Indeed, the amount of money in the form of grants from private industries and governments is already staggering.

As you consider the information in the next few chapters, keep in mind that if the so-called civilized and economically-developed countries enact laws outlawing certain experimental procedures because of the moral and ethical outcries of their people, those involved in this research and experimentation will move their laboratories to Third World countries, where they will be welcomed with open arms and no restrictions. *Again, there is no end to the combination of new and enhanced creatures that will be developed.* Use your imagination! A horse with a human brain—maybe with feet and hands instead of hooves—or any other human features the experimenter may desire, is not out of the question.

In 1965 Gordon E. Moore, co-founder of Intel, described a trend he had observed where the number of components in integrated circuits (IC's) had doubled every year, from the invention of the integrated circuit in 1958 to 1965. He predicted that the number of components that could be placed on an individual IC would double every two years. This became known as Moore's Law. He predicted that it would continue that way for ten years. In fact, the number of components since then has continued to double every 18 months to two years.

There is another law, Butler's Law, which says the amount of data coming out of an optical fiber doubles every nine months. All of our electronic and digital capabilities have been doubling every 18 months or so. It is now currently predicted to remain at that rate until 2020. This will be discussed in more detail in the coming chapters. All of this is to say that by the time you read this, much that is currently on the drawing boards may have already transpired. Technology is progressing that rapidly.

As we go through the scientific topics in this book, there will not be any credit given to any individual or laboratory team; this is written in general terms. However, if you desire such information on any of the topics or words, using a search engine on your computer will yield readily available information.

The definition of *convergence* used in this book is: *Individual objectives moving independently toward a point of focus or outcome; progress or technological achievement becoming more frequent and more intense as time goes on; and distance to outcome becoming shorter.*

The individual lines of convergence will gain momentum (referencing Moore's Law) if nothing occurs to slow them down. For example, this could mean an intensification and frequency of violence and corruption on Earth, which the Bible predicts (Luke 17:26, Genesis 6:11), or an increase in the scientific knowledge base of mankind, which the Bible also predicts (Daniel 12:9–10). We will explore these and others.

We begin the first chapter with the convergence of *Transhumanism*, its goals and ideals.

—**Lewis E. Hildreth**
July 2012

CHAPTER 1

TRANSHUMANISM

*T*RANSHUMANISM IS A big word which many people probably haven't heard of. It is the end product, the highest goal, of the converging sciences and technologies. It is that part of the equation which is 35 to 50 years or more in the future, where we are headed, if God should so allow. Perhaps some may wonder why this book would start with a topic that is farthest in the future. It is because this author doesn't want you, the reader, to "lose sight of the big picture," the end product, where these scientific achievements are all going, as we study the "pieces-parts" of scientific and technological advancement that will come together (or *converge*) to achieve the outcome of Transhumanism.

What is Transhumanism, which is sometimes referred to as post-humanism? According to its proponents within this international intellectual and cultural movement, it is defined as not only the *possibility* of fundamentally transforming the human race, but *the desirability of fundamentally transforming the human race.* They want to "evolve" the human species into something greater! Using the technologies that are now considered to be in their infancy, this intellectual community plans to eliminate aging and greatly enhance human physical and intellectual capacities. They see a future where all diseases such as colds, flu, and pneumonia will be completely eliminated. Diseases and viruses like AIDS, malaria, dysentery, and SARS, plus all other infectious diseases and physical ailments that have plagued humankind from the beginning, will no longer exist to strike human beings. Things like the destructive genetic mutations that cause most cancers (hereditary and non-hereditary types) will be reversed.

Humans have a brain that could be interfaced with computer technology, or enhanced by genetic manipulation, or both, that could allow us to have mental capabilities far beyond our imagination. We could possess a photographic memory, instant recall, or the ability to play any and all musical instruments without practice, with professional quality or even beyond professional as it is defined today. Physical enhancement will allow muscular strength and endurance well beyond what can be obtained from drugs and steroids; i.e., to be able to go for days without sleep, yet still be alert. It is evident that in some people's way of thinking, anything we could do to improve human abilities would be an improvement to our race.

Who wouldn't look forward to a future like this? Human suffering and the challenges it brings could be greatly, if not totally, eliminated. Human life could essentially be extended indefinitely.

It should be noted that not all people working in these fields of endeavor are looking this far into the future. Many are just looking forward to achieving the next step from many previous steps already accomplished. Building on the works and labors of many hours and years of previous colleagues, they're just concentrating on achieving the next small, and yet great, breakthrough. It can be called research just for the sake of research, exploration just for the sake of exploration, discovery just for the sake of discovery. It is exhilarating for those in research that in some small way they can contribute to the knowledge base of humankind.

Spiritually speaking, where is this community of intellectuals who wish to advance Transhumanism? It is reported that for the most part they are atheists, agnostics, or secular humanists, although some may attend the churches of liberal Christianity. They believe in evolution, that the human species evolved from some lower life form and that Transhumanism is just the natural extension of that evolution. Creationism is out of the question. From the vantage point of this spiritual worldview, there is no God—or if there is, He is entirely impotent. Given a viewpoint of "there is no God," then we are not subject to be accountable to God. Humankind is then free to make up its own ideals of right and wrong, their own concepts of morality and ethics.

God created us in His image. But if there is no God, as many claim today, then we are free to re-create life in any image we desire.

We have already started the re-creation of life. Transhumanism is already here, albeit in its infancy stage, through gene splicing, cloning, and brain-computer interface. It will continue as long as God permits it.

Notice and consider, however, what these intellectuals are endeavoring to deliver with "enhanced human abilities," such as long life, without aging or illness. These promises are the same as the promises of God, except they leave God out of the equation. Do they aspire to usurp the domain and province of God? It seems so.

God promises eternal life to those who believe: "… no more death or mourning or crying or pain" (Revelation 21:4). "The wolf will live with the lamb, the leopard will lie down with the goat, the calf and the lion and the yearling together; and a little child will lead them. The cow will feed with the bear, their young will lie down together and the lion will eat straw like the ox" (Isaiah 11:6–7). This sounds like God's version of genetic engineering!

How long will God permit the introduction and advancement of these coming scientific achievements? How far will He allow them to go? I do not know. However, we do know that, speaking of the coming tribulation in Revelation 9:6, God says that "… during those days men will seek death, but will not find it; they will long to die, but death will elude them." Will that perhaps be because men have been genetically-enhanced? Would they be mechanically and genetically manipulated to the point that what would kill ordinary men would only cause them suffering? I cannot say. It is speculation—food for thought.

How will all of this, what seems to be science fiction, transpire? We go back to the preface, about genetics, robotics, artificial intelligence, and nanotechnology. Another similar group is nanotechnology, biotechnology, information technology, and cognitive science. Nanotechnology comes up again here because it is a key that the other major fields depend on for their ability to advance. It is a major player in the "convergence" scenario. Nanotechnology will be discussed in greater detail later. These and other sciences are what Transhumanists are depending on to be the mechanisms or the tools that will radically alter the design of our minds, our physiology—and our future offspring, if we get that far. These technologies will fundamentally change the nature of human beings.

The U.S. Department of Defense has reportedly accelerated their research on brain and body alteration technologies. They want to create "super soldiers" that would give the United States distinct battlefield advantages. Visiting the U.S. Army Research Laboratory's website can be an eye-opening, educational experience. If this is public information, one must wonder what is going on behind closed doors.

Remember, the LORD said, *with a common language nothing would be impossible for them to do.*

We have been given a timeline that comes from the scientists themselves. It is their best guess and it might be optimistic. They base it on the past patterns of exponential increases in technological capabilities (Moore's Law). *Without the intervention of God, it is not "if" these things will occur, but "when."* There could—and probably will be—drawbacks and disadvantages to this coming technology. Some of these possible drawbacks will be pointed out in later chapters as the need arises.

Now you have had a peek into what Transhumanism will look like. What form will Transhumanists seek to adopt? It is too early to tell. There are competing theories as to how it might develop. The front-runners are humans in the form of *cyborgs* (part human and part computer/machine), genetically-enhanced humans, or a combination of both.

We will examine the case for cyborgs next.

CHAPTER 2

CYBORGS

UNLESS YOU'VE BEEN living under a rock, almost everyone is familiar with cyborgs. You just might not have realized it! "Captain's log, Stardate 2024. These are the voyages of the Starship Enterprise." Sound familiar? It certainly will to the so-called "Trekkies," the fans of the long-running "Star Trek" TV shows and movies. Or how about this: "A long, long time ago, in a galaxy far, far away...." This would be instantly recognizable for those who follow the Star Wars movie episodes. All of these science fiction movies are well-populated with cyborgs of all sizes and descriptions.

So, what is a cyborg? Good question; the answer depends on who you ask. With all the ways they have been portrayed on TV and in the movies, an exact definition may be hard to come by. Generally speaking, a cyborg is a "cybernetic" organism. In real life, it could be insect, animal, or human. It means a creature that is enhanced electronically, mechanically, robotically, or by a combination of those three in varying percentages.

Captain Picard of Star Trek fame almost became a fully functioning cyborg in one episode. Without going into great detail, he was captured by the cyborgs and the process of integrating a computer into his human body began. It was a wearable-type computer that had the ability to create itself on the human body at the same time it was integrating itself with the human brain and nervous system. When the process was complete, Captain Picard would irreversibly be a cyborg who would be controlled by a centralized computer system. He was rescued just in time!

How about Darth Vader from the Star Wars movies? He was a cyborg as well, with robotic and electronic enhancements. He was not controlled by an outside computer source, but instead had the ability to control the minds and actions of lesser, intelligently—and electronically—inferior beings.

Science fiction can dream up all kinds of cybernetic creatures endowed with all manner of abilities and capabilities. It is fun to pretend, to make believe, limited only by your imagination. But remember, God said that "nothing they plan to do, [nothing they can imagine] will be impossible for them" (Genesis 11:6)—if they have one common language.

The airplane was a major new technological innovation that came about in the early 1900s. Airplanes continued to be developed and advanced throughout the wars of the 20th century, beginning with World War I. Pilots of every era trained to fly these airplanes, either in actual combat or simulated combat. Almost every pilot developed a love affair with his aerobatic steed. They became "one with the airplane." It became a part of them psychologically. The fighter aircraft was just an extension of them. They strapped the airplane on; they wore it. While man and machine developed an intimate relationship, it was a strictly one-sided relationship. Yes, they were using the latest technology, but the technology was not using them. They were not cyborgs.

Where is the technology of cybernetic development today? How far have we come and how far do we have to go? This information will no doubt be outdated by the time many read this. Given the military's keen interest in this technology, we will start with the "cyborgification" of the insect and animal kingdom.

"Bug Borgs" ... has a cute ring to it, doesn't it? When I was a kid living in the country, my friends and I would sometimes capture June bugs. Having captured them, we would tie a string around one of their hind legs and release them. They would fly around in circles at the end of the string until they (or we) tired of it. We would then take the string off and allow the June bug to go on its merry way. I suppose this may sound cruel to some, but it's nothing compared to the experiments that are being done in the name of "science" and "technology" today.

How would you like to have an insect, say a bee, which you could control from a remote position? Besides controlling the bee, you could implant sensors into the bee (or bees) during the pupal stage. Depending on the sensor or sensors, you could use the bee to detect drugs, explosives, gas fumes, or chemicals; have a microphone in it; or maybe even have video surveillance capabilities. This can all be controlled by microelectronic mechanical systems. Like a remote-controlled airplane or helicopter, many types of insects could be caused to fly, hover, or land as long as it was in range of the controller.

Since this technology can be implanted into insects, the same technology can also be implanted into the animal kingdom. Birds, dolphins, sharks (you name the animal), it will be done and for the same reasons. Maybe you wouldn't want an insect like this, but the military, homeland security, and certain police forces certainly would. This advancement in technology is largely funded by the military. Like so much of what the military has in the way of guarded secrets, the details of much of what we have discussed here would be considered "top-secret." Thus we do not ultimately know how advanced the technology is.

Remotely controlling flying insects has already been demonstrated. Are we humans next? As a popular saying goes, you betcha!

The short definition for cyborg is a being with both biological and artificial parts. As defined earlier, its artificial parts could be solely electronic or electronic with mechanical and robotic enhancements. Surprise ... Grandma could be a cyborg! As mentioned at the beginning of the chapter, almost everyone is familiar with cyborgs. Assuming you somewhat know your grandma, how could she be a cyborg? Well, if she has a pacemaker or maybe a cochlear implant, that technically (and by definition) is a cyborg. She would be a biological being with artificial electronic parts.

The pacemaker monitors electrical signals from the nervous system, maintaining the natural rhythm of the heart.

If that rhythm should become irregular, the pacemaker will transmit an electrical signal to re-establish the natural rhythm. In the same manner, a cochlear implant as an aid to the hearing-impaired will transform sound vibrations into electrical impulses and transmit them to the cochlear area of the brain, where the impulses can be interpreted as sound. These implantable devices and others like them are considered "restorative" in nature. They restore back to normal (or close to normal) functions of the body that have been lost or impaired. This would include organs and some prosthetic limbs.

No one would begrudge Grandma her pacemaker. It is advancements in restorative technologies that enhance the quality of human life. In the next chapter we will discuss how the restorative qualities of genetic engineering have the same effect on the quality of life. This isn't the area where red flags are being raised. The red flags are being raised in the area of human and animal enhancements.

Human enhancements go beyond restoring bodily functions to normal. The goal of enhancements is optimal performance: gaining new bodily functions and enhancements that were not originally present in humans, maximizing human output. The TV series of the 1970s known as "The Six Million Dollar Man" (a.k.a. the bionic man) can be an example. In case you haven't viewed that program, Lee Majors, the star of the show, was in a terrible accident. He was important to the government, so they decided to rebuild him "better than new." He could run 60 mph, lift cars by himself, and do other superhuman physical feats. Although the term cyborg was probably not in common usage then, he was somewhat of an example of what cyborgs of the future would be. The technology is quickly developing to make these sci-fi characters reality. It is technology that is converging with other sciences and technologies, heading us in the direction of Transhumanism.

Moore's Law has been mentioned, regarding how technology is doubling every 18 months. This is also known as exponential growth. Exponential growth can be very deceptive because it starts out very small, with hardly any noticeable growth. Then, before you know it, it snowballs into something that may even be uncontrollable. Let's look at an example of how it works.

There was a young man who was looking for work. He was asking his prospective employers for $20 an hour. Even though he was worth twice that, no one would meet his request. He decided to change his tactics. The next interview he had was with a greedy boss of a mid-size company. The boss offered the young man five dollars an hour for 30 days. Because of his need for a job, the young man (knowing that he was being taken advantage of) decided he had nothing to lose by offering the boss a deal. He told the man, "If you will sign a contract with me, I will work for you the first day for one cent. The next day you will double my wages to 2 cents and double it every day until the 30 days are up." The greedy boss took him up on it. "What employer could refuse a deal like that?" the boss thought. The young man's daily wage at the end of the first week was 16 cents. At the end of the second week his daily wage was $5.12. So far the boss was getting a great deal! But let's jump ahead to the last day, the 30th day. The young man's wage on this

30th day was $5,368,708.80. That's exponential growth! And that's how it can sneak up on you. Don't believe it? Do the math. This is how the growth will occur in science and technology, if left unchecked.

If your great-great-grandparents who died in the 1800s could come alive today, they would be absolutely dumbfounded by what has taken place since their era. They had no refrigerators, no freezers, no radios, no TV's, and no calculators. There were no automobiles, no airplanes, no way of dreaming about putting a man on the moon. As awed as they would be, it's nothing compared to the changes we could see in our great-great-grandchildren's lives.

To know what's going to happen with the future technology, it would help if we knew what was happening now. Without going into great detail, there have been successful brain/computer interface (BCI) demonstrations where the computer reads our brain waves, our thoughts, to enable the computer to control external devices. It is like voice recognition software without the voice. It can be used in many different ways to the benefit of mankind, like controlling a wheelchair by thought. Controlling a prosthetic forearm and hand by thought is reality, on an experimental basis. Thought-reading and thought-transference between two humans via computer has been demonstrated. This tells us that soon we will be controlling external devices by thought processes alone. Be careful! If thought/brain waves can be received by external devices, then those devices can receive those same waves and read your mind. *There will be no secrets.*

The future ... what's it all about? Well, just use your imagination. *God said there is nothing we can't accomplish with a common language.* This author believes a big part of "what it's all about" is *control*. If you can control things telepathically, then they can control you. Who is the "they?" I do not know. However, we do know that with human nature being what it is, some "one" or "thing" will be there, doing the controlling. Ultimately, it *will* be the Antichrist, or a world dictator with the same power as ascribed to the Antichrist. One thing about the science fiction cyborgs is that most were controlled by a centralized computer/intelligence, keeping them in lockstep with whatever was deemed politically correct and expedient.

"He also forced everyone, small and great, rich and poor, free and slave, to receive a mark on his right hand or on his forehead, so that no one could buy or sell unless he had the mark which is the name of the beast or the number of his name" (Revelation 13:16–17).

Centralized control *will* happen in the future, because it *has* happened in the past and *is* happening now. *Overwhelmingly, the population of the Earth was and is unaware of this control being exerted.* Note this passage from the book of Job, Chapter 1, starting with verse 8:

> Then the LORD said to Satan, "Have you considered My servant Job? There is no one like him on earth; he is blameless and upright, a man who fears God and shuns evil." "Does Job fear God for nothing?" Satan replied. "Have You not put a hedge around him and his household and everything he has? ... But stretch out Your

hand and strike everything he has, he will surely curse You to Your face." The LORD said to Satan, "very well, then, everything he has is in your hands, but on the man himself do not lay a finger." ... One day when Job's sons and daughters were feasting ... a messenger came to Job and said, "The oxen were plowing and the donkeys were grazing nearby and the Sabeans attacked and carried them off. They put the servants to the sword and I am the only one who has escaped to tell you!" ... While he was still speaking, another messenger came and said, "The Chaldeans formed three raiding parties and swept down on your camels and carried them off."

The point to be made here is that the Sabeans and the Chaldeans had no idea who was behind their thoughts and plans. They didn't have a clue that they were being manipulated from an outside, *extra-dimensional* force. As far as they knew, it was their own ideas and a sure way to gain some wealth. Individuals, groups of people, even whole nations have been influenced—and *are* being influenced—by this evil, centralized, controlling force we know as Satan, and they don't even know it is happening.

Be careful about scoffing too quickly at this idea of an extra-dimensional entity being able to control our thoughts. Some might want to account for this phenomenon in a more natural way, like labeling it as "emotional contagion," or some other subliminal psychological effect. As a suggestion, at least be open to the possibility of "other dimensional" interference and influence in individual lives, as well as national tendencies. *Is* the direction our world is going in being satanically controlled from another dimension? This author believes it to be so.

Through genetically-enhanced brains, the power of the brain interfaced with computer technology will be more formidable and controllable than ever. The next chapter will examine genetically-enhanced humans and animals.

CHAPTER 3

GENETIC ENGINEERING

O F ALL THE chapters presented so far in this book, this one on genetic engineering is by far the most complex, the most far-reaching, and for me, personally, the scariest. I am not a "technophobe" (a person afraid of technology), but the ramifications and potential of this technology—both for good and for evil—have almost no bounds. Under the heading of genetic engineering are subheadings of biotechnology, DNA research, gene splicing, synthetic biology, artificial life, and more. I will try to define and present where each subfield presently is, technologically speaking, and where they may be headed in the future. These fields are so interrelated that it is hard to know where to start to have it make sense. Perhaps the best place to start is in the beginning, with some basic information.

An engineer is a person who is trained and educated to be able to design, test, and build things. A mechanical engineer might build or improve buildings, automobiles, and such. In the same manner there are also electrical engineers, electronic engineers, chemical engineers ... you get the picture.

Humans have been genetic engineers ever since they learned to domesticate plants and animals. For example, they found out that by crossbreeding certain varieties of the same food crops, they could enhance that particular plant species, allowing it to possibly produce more food, be stronger, heat-resistant, and so forth. They also learned early on to do the same thing with domestic animals. Of course domestic animals were not "domesticated" until they had some primitive genetic engineering performed on them. Early man could get "beefier" cows or more tame cows or pigs, or better hunting dogs, just by interbreeding like-kind animals that had traits desirable to humans.

It is a human-forced form of micro-evolution, known as speciation, where traits and attributes can be enhanced *within* a species. For example, a dog can be bred to have certain traits or abilities enhanced over other dogs. But it always remains a dog and cannot be forced to become a different animal. This has been the method for thousands of years, right up until the early 1980s. This was a very slow process and indeed sometimes took many, many generations to see the desired results.

However, in the 1980s things began to change. We began to find new ways of genetically-enhancing plants and animals, for the same reasons and purposes our ancient ancestors did it. These new ways produced results exponentially faster than anything previously conceived. The age of genetic engineering had begun in earnest and continues today. Now, obviously, there was a lot of research by many people and laboratories in the two or three decades leading up to the 1980s to reach this point of learning how to change a plant genetically. It didn't just happen. Without going into the individual steps, suffice it to say that certainly we have come a long way since the 1980s.

Whether you like it or not, you are eating genetically-modified foods. Even the meat we eat has been raised on genetically-modified crops. One report shows that as much as 95 percent of our corn crop has been genetically-modified, along with soybeans, tomatoes, potatoes, and even tobacco (though not a food crop), just to name a few. As of 2009, eleven genetically-modified food crops were grown commercially in 25 countries, with the United States being the largest producer.

My nephew farms large acreages of ground, mainly planted in corn, wheat, and soybeans. I asked him recently about these genetically-modified crops. He told me that you could still get ordinary hybrid corn, for example, but for the most part farmers were planting "GMO's." And what is a GMO? It is a "genetically-modified organism." He told me in corn, for example, that there are ten levels of modification. Depending upon the ground, the climate, and other extenuating circumstances, a farmer could order and plant GMO corn tailor-suited for the local environmental conditions. Farmers want their ground to produce the most crops possible, the most "bang for the buck." Some of these levels of modification include being resistant to certain herbicides. After all, when a farmer sprays his fields to kill weeds, he doesn't want to kill his corn too. Another level allows the corn to produce an insecticide which kills or stops various types of bugs that normally will damage the crop. Some levels are drought-resistant, cold- or frost-resistant, fast-growing for short seasons, and so on. This all sounds beneficial to mankind, right? But just how is eating this insecticide-producing corn affecting those who eat it?

There are three groups of genetically-modified crops. The first group allows the plants to be resistant to insects and herbicides in order to increase crop production. The second group is modified to tolerate cold, drought, heat, and salt. The third group that has been genetically-modified is called pharmaceutical crops. They produce various drugs and vaccines. All of this is possible through genetic engineering. They are able to do it by splicing and combining genes of different plant species or injecting bacteria genes and even insect and/or animal genes into the genetic makeup of the plants, in order to obtain the desired results.

Are you starting to see the things that are *converging* toward "something," just in the scientific and genetic realms alone?

Many centuries before Christ, mankind figured out how to manipulate bacteria, mold, and yeasts. They may not have known the specific players involved in the game, but they did see the results of their activity. The early Egyptians could brew beer and get bread to rise. They soon found a way to process spoiled milk into cheese.

Today through gene splicing (a form of genetic engineering), untold thousands of applications are the result of these genetic combinations. An example of what genetic engineering has accomplished through redesigning bacteria, yeasts, and mold is oil-eating bacteria to clean up oil spills and other toxic wastes. Methane gas, proteins, enzymes, and alcohol-producing bacteria are in production and use, with many having commercial application. The production of myriad types of proteins and enzymes is what has allowed gene splicing to exponentially increase.

As an example, the genes being spliced into bacteria are not limited to other forms of bacteria. They can be genes from any other living species. The bacteria then become little "factories" that can produce medicines of all kinds, as well as other by-products useful in so many areas of manufacturing. They have application in agriculture, the fuels industry, the medical industry, and research. Genetically-enhanced bacteria can be used as "information storage units," since you can implant various genes (genes contain the genetic information controlling certain traits of the organism) into bacteria for use at a later time. This is usually done by freezing the bacteria, then retrieving the genetic information by thawing the bacteria at a later date, when needed. The genetic combining of two or more organisms is known as *recombinant DNA*.

Without going into great detail about viruses, note that scientific laboratories have the same ability to manipulate the genetic information in viruses, to mix and match genes with any living species to achieve desired outcomes. We can develop medical immunities against certain viruses, as well as create viruses where no human immunity exists. This obviously has implications for military application.

I do not want to imply that no further research is needed because "science has all the answers." However, I do state that genetic engineers have come a long way. There is still much more to be learned; it is just a matter of time and money for research. Meanwhile, let us hope that none of these new "little guys" (products of research thus far) don't escape from the laboratory!

The definition of genetic engineering is: the use of DNA technology (genes) to insert the natural DNA or synthetic (man-made) DNA into any living species, supplementing or replacing the target organism's DNA, for the purpose of obtaining a result or outcome desired by the laboratory person or team. Any organism created this way is considered a genetically-modified organism, a GMO.

At this point, let's clarify the two types (or purposes) of genetic engineering. One type is called gene therapy and is used mostly in and for humans. This therapy is used to restore lost function. This type of genetic enhancement is *not* hereditary. It cannot be passed along to offspring.

The other type of genetic enhancement involves inserting genes/DNA into the host embryo. This is mostly a partial supplement of the host's DNA and permanently changes the target host organism *and its offspring indefinitely*. These changes are referred to as "germ line" changes and they *are* hereditary. Theoretically, this applies to all living organisms, including humans.

There is a picture that I kept from the April 2006 issue of *Popular Mechanics*. It is entitled "Green Pigs and Ham." The explanation with the picture states: "Taiwanese scientists injected jellyfish DNA into pig embryos to produce three fluorescent-green, glow-in-the-dark piglets. Claiming that all of the pigs' cells are green, the scientists argue that the pigs are useful: when their stem cells are implanted in other organisms, the fluorescent green will be easy for researchers to track." Anybody want a ham sandwich? If this is how advanced the technology was in early 2006, one can only imagine how advanced it is today when one considers exponential growth.

This genetic activity could and *will* include humans. It has so far been used for the good of mankind, *as far as we know*. But it *can* also be used for evil, for the total transformation (and possibly the destruction) of mankind.

Though I have no real proof, I sense and believe that genetically-enhanced humans of some type likely already exist. These things would have to be kept absolutely secret, awaiting the right timing to be revealed. Genetic engineering is advancing very quickly and seems to be outpacing some of the other scientific fields, such as cybernetics.

Each cell of your body contains DNA. DNA is only an "information storage system." There is no *functional* difference between your computer hard drive, CD, or any other information storage device related to your computer and DNA. They both function as storage systems. A DNA programmer, i.e., genetic engineer, does with DNA much the same as a computer programmer does with a computer. A genetic engineer in essence is involved with moving or exchanging genetic information from one organism to another. (This explanation is not meant to minimize the effort and education required to be involved in this cutting-edge technology. The requirements are phenomenal.)

There is more information in each cell of your body than what is contained in all volumes of a popular encyclopedia, when that encyclopedic information is converted to digital bytes. From one cell, your body can *theoretically* be reproduced.

Under the heading of genetic engineering is the subheading of *cloning*. Presently on the Professional Bull Riders' circuit (known as the "PBR"), there are four bucking bulls—performing quite well in that sport—that have been cloned from a single, championship-caliber bucking bull. A champion horse used for barrel racing has been cloned. And of course there was Dolly, the sheep. If science can do that with such complex animals, then what else can they do or have they already done?

Are you starting to see what is meant by things *converging*? Where does it end? It has been reported that a bear that had been dead four years has been cloned back to life. Good, healthy DNA was retrieved from the dead bear, then inserted into the embryonic cell of the host mother and brought to term. This new little bear was genetically-identical to the deceased bear.

The Japanese announced recently that they would try performing the same thing with a 10,000-year-old woolly mammoth. If and when they are successful, then what is next—a so-called 30,000-year-old Neanderthal? They already have DNA samples from Neanderthals, trying to see if Neanderthals are genetically-related to modern humans. If they can do that, perhaps it will not be the Japanese, but *somebody* will clone a Neanderthal. Anybody want to recreate "Jurassic Park" dinosaurs?

Superman, the popular and fictional action-hero of the 1950s (and beyond), was "stronger than a locomotive, able

GENETIC ENGINEERING • 25

to leap tall buildings with a single bound." He fought for "truth, justice and the American way." Superman was said to have come from another planet. Outwardly he looked human. Inwardly, he was obviously genetically-modified and enhanced, but nobody realized that at the time. Genetic enhancement in the 1950s was still just science fiction. Superman was *originally* moral, just, had a keen sense of right and wrong, and used his superhuman abilities for the good of mankind. Had the person who was manipulating his genetic makeup been up to no good, then he could have just as easily given Superman the desire to kill, steal, and destroy.

Genetic engineering has the ability to change the appearance of humans, to increase intelligence, change behavior, change adaptability, change endurance, change their character—and everything else about us—forever. Many fear and work to prevent this outcome. However, the Transhumanists look with favor upon the desirability of these changes. Who do you think will win out? While these abilities may yet be 35 to 50 years in the future, what will the experiments look like in the meantime? They will use this technology on animals first. They will change the animals in the same ways as just listed for changing humans. They call it "animal-enhancement" or "animal upgrading." The dog referred to earlier in this chapter can now be forced to take on traits and features of different animal species. How smart do you want your dog to be? Do you want it to be self-aware or have the ability to speak words? It could be a canine/leopard hybrid or canine/human/goat hybrid or canine/_____ *you* fill in the blank! The paths these experiments take are limited only by human imagination. Some will no doubt be very grotesque. The "Chimeras" of ancient civilizations are not out of the question. (A Chimera was a mythological, fire-breathing monster, commonly represented with a lion's head, a goat's body, and a serpent's tail.)

Creation is beginning to be radically changed. However, I personally believe the second coming of Jesus Christ will occur before these things find total fruition.

If you have read the history of Germany during World War II and the experimentations conducted on humans, as sanctioned by Adolf Hitler, you know Hitler and his medical doctors were all about finding ways to genetically-enhance his "Aryan" race. They ultimately were limited by the "old-fashioned" way of genetic engineering. One question: What do you think Hitler and his cronies would have done to animals and specifically to humans, had they been in possession of the knowledge we have today? We must shudder to think what they would have done! Let us hope and pray that another type of Hitler *never* comes into power again, because that person would make a "Superman" that would be totally controllable, with the built-in desire to kill, steal, and destroy. Yet, we *do* know a greater "Hitler," the Antichrist, *will* come on the scene in the last days.

Thirty-five to fifty years in the future, these things are predicted to come to total fruition. Some of you will say, "Thirty-five to fifty years from now is well past my life expectancy—it's a long way off. It doesn't affect me, so why should I concern myself with it? Maybe it will affect my children and grandchildren, but they can deal with it then."

I call this the "Hezekiah Effect." Hezekiah, king of Israel in Old Testament times, showed the emissaries from Babylon all the gold that was contained in the palace treasury. It is possible he even showed them the Temple. The inside of the Temple walls, the floor, the ceiling, and the fixtures were solid gold. Hezekiah told Isaiah the prophet that there was nothing in his entire kingdom that he did not show them. Isaiah told the king that the time would come when everything in his palace, including his descendants, his own flesh and blood, would be carried off to Babylon. According to 2 Kings 20:19, Hezekiah responded to Isaiah: "'The Word of the LORD you have spoken is good,' Hezekiah replied. For he thought, 'Will there not be peace and security in my lifetime?'" It didn't affect him, so he wasn't going to worry about it.

The information that has been presented thus far can certainly be unnerving. It is not meant to scare you, but to help you put more trust in Jesus, to help you see the great need for God to intervene in this world system of things, to help you pray more fervently for the soon return of Jesus and the setting up of His kingdom on Earth.

After the next few chapters on the topic of converging fields of science, we will then examine other "emerging" factors that could have a bearing on whether the optimistic scientists and Transhumanists will have the time needed to accomplish their goals. The next topic for consideration is artificial life.

CHAPTER 4

SYNTHETIC LIFE

CALL IT SYNTHETIC life or artificial life, the name implies what it means: man-made life or life as we *never thought it could exist*. This is not a scientific paper. It is not designed to be read by the researchers described herein. The goal is to present a logical and understandable format that conveys basic information to the "scientific novice," enabling him or her to see the ramifications of the bigger, broader picture.

Synthetic or artificial, life science is a subheading of genetic engineering. While most genetic engineers are interested in the recombining of DNA from various organisms, synthetic life scientists are mostly interested in producing synthetic and artificial DNA to insert into already living organisms, or else to create a totally synthetic life form.

The synthetic life engineers generally fall into two camps, known as "top-down" or "bottom-up." Top-down engineers will take an already existing organism, remove its DNA (genome), and insert synthetic DNA in its place. They still use the cell membrane and cytoplasm (the gel-like substance residing between the cell membrane, holding all the cell's internal sub-structures, except for the nucleus) of the existing cell. The bottom-up engineers start from scratch and attempt to build a completely artificial, synthetic cell. There will be more on this shortly.

One of the ways I like to start "beginning of life" conversations with church youth groups is by taking a mousetrap into the classroom. After snapping the mousetrap a couple of times (which results in a loud crack), I usually have their attention. I then pass the mousetrap around (requesting they be careful with their fingers) and ask them how many individual components they can count that make that mousetrap function. Most come back with the answer of nine, which on this particular mousetrap is correct. I then draw their attention to the fact that each of these nine components is different from the others. Each is made with different design characteristics and has different functions, all working together doing their own jobs, to allow the total mechanism to do what it was "designed" to do: kill mice. I then ask, "What would happen if we removed any single one of our choice of the nine components?" Well, this is an easy conclusion for them: the mousetrap will not work. It will not kill mice. I then explain to the young people that the mousetrap is "irreducibly-complex." It has "minimal" complexity, meaning that the removal of one or more of its components renders it useless as a mousetrap.

I then explain to the young folks how I came into possession of this mousetrap. You see, I was walking along in an open field and a thunderstorm was coming up. I saw a lightning bolt strike about 100 yards in front of me. Boy, was that scary! I went over to examine the ground where the lightning struck and there I found the mousetrap. The electrical energy from the lightning, plus the heat, took the chemicals and minerals in the ground, along with the influence of an electromagnetic pulse generated by the lightning bolt, and *created* this mousetrap.

Can you hear their response? I just lost all credibility with them. They *instinctively knew* that there was way too much design and engineering input for this to have been a random act of nature! Try as I might, I could not convince them otherwise. They said they would have to be crazy to believe me. I then point out that if they have detected a complex design, there must be a complex designer. Next, I go on to explain irreducibly-complex, minimally-complex cells to them.

Like the irreducibly-complex mousetrap where you could add features to the mousetrap and it could still function, taking it below the irreducibly-complex minimum causes it to cease to function. So it is with an irreducibly-complex cell. The simplest cell has around 400 gene products. The implication is that if just one gene product is removed, the cell ceases to function. The artificial life engineers have also discovered that it takes hundreds (if not thousands) of different proteins, enzymes, and chemicals for a cell to function.

But what about a "complex designer"? The information presented in the next several paragraphs is, I believe, empirical evidence for a Creator God.

Most genetic engineers (and especially synthetic life engineers) believe all living things have been made from non-living chemicals. The right chemicals, whether on the ground or in the air, with the right temperatures and maybe the exactly right electrical discharge that existed in the "primordial soup" of early, primitive planet Earth, all came together by *accident* to produce the first living cell, from which all life evolved. Thus, creationism and the idea of a Creator are out of the question. The synthetic life engineers are so convinced that the first cell of life started from the random assembly of non-living, inert chemicals that they think it's laughable to consider any other possibility.

In the laboratory, scientists try to duplicate the process that took place "by accident" eons ago on planet Earth. Through nanotechnology, they have created many, many new chemicals, in the form of proteins, enzymes, and such. With these new chemicals, even ordinary genetic engineers can use these tools to cut and splice, add or delete DNA and the DNA fragments of their choosing to produce the kind of living organism they are shooting for. With this new technology they are able to control every aspect of this process. Checking the outcome of each experiment, the scientists can then evaluate the results. They can then change the types of chemicals and their percentages and again observe the outcome.

Here is what the scientists need to do. The bottom-up synthetic scientists need to create *from scratch* the DNA strands that will control their organism, and in this case, a single cell. In order to do this, they need hundreds (if not thousands) of protein and enzyme types manipulated by monomers and polymers. Each protein and enzyme that makes up, or helps form the genes, is chemical in its composition. They need to be chemically different from each other, each with different functions, and they need to be placed in the proper order. With the thousands upon thousands of combinations in which these chemicals can align with each other, the exact alignment of the DNA base pairs must happen. Some genes regulate other genes. The regulated genes in turn regulate others. They must all come together simultaneously, in the very same moment, or the cell doesn't work. This process is painstakingly time-consuming and requires the most highly-educated people imaginable.

Next, a cell membrane is needed. This membrane is also made of chemicals that have to do several things. First, the membrane has to contain (keep from escaping) the DNA pairs and cytoplasm that are found inside any normal cell. This membrane must also protect its cargo of DNA and cytoplasm, among other things, from the surrounding environment. From the surrounding environment, that cell membrane must allow the proper food and energy—which is needed by the cell to fuel it—to pass into the cell proper, while at the same time keep out ingredients harmful to the cell. Once the cell processes the food and energy it received from outside the environment, it must be able to pass the by-products of cell metabolism back through the membrane into the surrounding environment.

Now, if the synthetic life scientists can accomplish all of this, they are still not home yet. If this cell is going to be a truly living organism, it must be able to pass on 100 percent of its genetic code to its offspring. *It must be able to reproduce.* Self-replication is another extremely high hurdle to get over. All of these hundreds and even thousands of very complex processes must come together simultaneously for this to work. *It could not have evolved slowly to this point.* It could *possibly* have evolved somewhat *after* it all came together, as in speciation, *but the first living cell had to come together instantly.*

As you can see, this is quite complex and time-consuming. It requires years of trial and error on the part of scientists to get every little aspect right. Here are some things I would like for you to consider about this process: It is all done in a laboratory. The laboratory is spotless and uncontaminated. Every aspect of this research is controlled: the temperature of the room, the humidity, the absolute sterility of the atmosphere; men and women are working in an absolutely clean and scrubbed environment. All of the hundreds of chemicals are individually stored. The chemicals are controlled and manipulated by highly-advanced equipment and highly-intelligent people. It takes all of this to even come *close* to producing that first artificial organism.

Now let's consider primitive Earth. To start with, it was a very hostile environment for life to "self-assemble." According to scientists, primitive Earth had no way to separate and segregate the chemicals. The "primordial soup" where life was supposed to have started was highly-contaminated and toxic. They know from their laboratory experience that some of the chemical processes needed to produce one aspect of the living organism were also the same processes that would kill other aspects. Yet according to these researchers, there was no "intelligent hand" guiding the needed events.

If man creates life, what's to stop him from creating something far more powerful and harmful than himself?

Scientists are on the verge of creating new man-made life in the laboratory. They can prove that by bringing the right chemicals, in the right order, at the right temperature, in the right amounts, and *in the proper sequences* (among millions of possible sequences), that life *can* be assembled from lifeless chemicals. They either do not believe in God or else they believe He is irrelevant, not having anything to do with creation.

However, I contend that life could *not* have started on this planet (or any other planet) without the direct involvement of an intelligent Being. I contend that primitive Earth and the primordial soup was so contaminated that life could *not* have originated by accident. Even with the work of highly-intelligent humans in a highly-controlled laboratory environment, creating life in a test tube has escaped achievement for the past 60 years. Keep in mind that the primitive, contaminated, and mindless planet Earth had to cause all of these hundreds of individual proteins, enzymes, and other, myriad nano-measures of individual chemicals to simultaneously and in segregated order *instantly* assemble and coalesce into one cell at the microscopic level, supposedly creating the first cell of life.

I believe this is empirical evidence that there *had* to be a Creator to put life in all of its diversity and complexity into reality. Every individual gene in a cell (all 400-plus of them) is just as unique and intelligently-designed as each of the nine components making up the mousetrap, and even more so.

The kids didn't believe the mousetrap came into existence by accident and yet so many of the genetic engineers want us to believe that the first living cell of life—which is infinitely more complex than a mousetrap—came into existence by "chance happening," by accident.

Of course, my beliefs wouldn't count with the genetic engineers! However, there are many researchers who have worked in the laboratories who have looked at the overwhelming evidence for a Creator and became believers in the God of the Bible *because* of the complexity and diversity they have discovered.

Colossians 1:15–17 says this about Jesus: "He is the image of the invisible God, the firstborn over all creation. For by Him all things were created: things in heaven and on earth, visible and invisible, whether thrones or powers or rulers or authorities; all things were created by Him and for Him. He is before all things and in Him all things hold together."

"In the beginning was the Word and the Word was with God, and the Word was God. He was with God in the beginning. Through Him all things were made; without Him nothing was made that has been made. In Him was life, and that life was the light of men" (John 1:1–4).

The many life-restoring benefits and the process of making life easier for humans with this research cannot be denied. However, as with all genetic engineering, there is *absolutely* a dark side—possible and probable.

Developing synthetic life has the same goals as all genetic engineering, using many of the same processes. The reason for highlighting synthetic engineering in this chapter is to show the absolute impossibility of life being able to self-assemble by chance happening.

God is the only reasonable answer.

CHAPTER 5

INFORMATION TECHNOLOGY

INFORMATION TECHNOLOGY, LIKE all sciences, has come a long way since its beginning. Back in the 1980s when corporations began embracing the computer in large numbers, there was then a need to add information technology ("IT") departments. The most visible component of such a department was the "help desk." For many employees, the new desktop computer at work was their first experience ever with a computer. They needed to take classes on how the various operating systems and software worked. There were also self-help "for dummies" books, all in an attempt to teach people how to operate this new technological device.

Before computer technology evolved to the point that allowed for desktop computers, most of the major universities and corporate laboratories had what were called "mainframe" computers. At this time the IT professionals were just starting to cut their teeth, so to speak, in developing and making these things work. Under the heading of information technology were subheadings such as computer hardware, developing programming languages, designing complex computer networks, and developing "hard drives" for information databases and storage.

The more IT professionals developed the computer technologies, the more it was realized that the range of knowledge needed to operate the systems was too much for any one person. Delegation of authority and specialization in specific computer fields led to college degrees in various related areas, such as data management, database and software design, engineering of computer hardware, networking, computer science, and programming, along with management and administration of entire systems.

Here are some examples, from personal experience, of how computer technology has advanced since the early 1980s. I was employed by a research and development laboratory. This laboratory had room after room full of mainframe computers. Each computer consisted of seven or eight cabinets that were two feet by two feet square and approximately 6½ feet tall. This lineup would be *one* computer. Each room would have a dozen or more of these lines. Each computer lineup had 12 hard drives for storage of information. Each of these hard drives was approximately 9 inches square and almost two feet long. They weighed about 100 pounds each and were capable of storing 375K of information (375,000 bytes of data). That is 4.5 megs, or 4.5 million bytes, of information per computer. However, this is "Stone Age" computing in comparison to today's megabillion storage capacities of desktop and laptop computers. Just one desktop computer today is more powerful than

INFORMATION TECHNOLOGY • 41

a building full of mainframe computers in the 1980s. If desktop computers are this powerful, can you imagine what the newest mainframe computers are capable of?

We have spoken in earlier chapters about exponential increases in technology, how technology—especially in computer science—has doubled every 18 months. There is a physical limit at which these exponential increases must come to an end. Most computer scientists think this limit will be reached somewhere around the year 2020. By that time integrated circuits, aided with the converging field of nanotechnology, will be operating, processing and storing information on the atomic (and maybe even subatomic) level. Computing power will be magnitudes of powers beyond what is available today in size, capacities, capabilities, and processing speed.

For what purposes could this power be utilized? Countless factories all over the world use computerized robotics in their manufacturing processes. Automobile manufacturers use robotic welders to spot-weld the metal parts of an automobile. Robotic insertion machines are used to automatically insert electronic components of all types into circuit boards. These insertion machines operate on an X/Y scale and are accurate to within 1/10 thousandths of an inch.

Most new jet fighter aircraft are called "fly by wire" aircraft. They have computerized flight control systems. The flight characteristics of these aircraft are so intricate and precarious that they would be uncontrollable by ordinary human pilot input. The new F-22 Raptor is reported to have three computers installed behind the pilot, allowing him to maintain control of this stealth aircraft.

If this is today, what's in store for tomorrow? Well, "pilotless" aircraft, for one thing. This is not a drone, which is a pilotless aircraft that is controlled by a human on the ground. No, this is a completely self-contained, self-programmed aircraft. We should not be surprised at this type of aircraft capability, because the same technology that would be used in that kind of aircraft is already being utilized by our cruise missile technology. The cruise missile flies low, recognizes the terrain over which it is flying, can hone in on any pre-programmed target using GPS technology, and yet be no further than three feet off-target. Pilotless aircraft are in the experimental stage now.

If you happen to be a fan of the Star Trek series, you will recall a robot by the name of Data. Data looked very human. He could process information and make calculations faster than any human on the starship. He lacked one thing: emotion. However, in one episode he was given a "memory upgrade" that allowed him to experience human emotion. The amount of computer power required to operate a robot like Data with total recall and with a fully-functional body is beyond our current capability. The Japanese have been leading in humanoid robotics. They have one that can stand upright and walk, as well as perform simple functions.

Someday in the not-too-distant future, we *will* be able to build Data. Will Data be self-aware like humans? Be able to know who he is and what he is? Will he be programmed to learn from his experiences? Will he be able to adapt to his environment ... always learning and never forgetting? Will he have the ability to procreate? Certainly not in the way humans do, but if computers today can assemble other computers (as just mentioned regarding computerized insertion machines), who is to say what they will do? And if they *can* self-assemble themselves, are self-aware, and self-programming, they *might* try to write their own future destiny. Could they not decide to take over and eliminate humans? Be careful ... they *used* to say we would never land a man on the moon!

We already have computers that can read and understand video game manuals, with the capability to construct strategies for winning, and this *without* human input. They recognize and know extremely well how to play the game the very first time. They then perfect their game-winning abilities by learning from their mistakes. Supercomputers now have the ability to emulate the output of the human brain. The vast majority of people do not understand what a "supercomputer" we humans have in the form of our vastly under-performing brain, meaning we only utilize about 10 percent of our brain's capability. The next step, according to the experts, is to give the supercomputer self-awareness.

Chapter 2 was about cyborgs—about integrating wearable computers with genetically-enhanced humans to produce the "ultimate" in enhanced humans, in what is considered the "natural evolution" of our species. This human/computer interface will come about due in part to advances in information technology and computer engineering. Let us look at how these advances will come about.

Buy a computer and part of that purchase price goes toward research and development. By purchasing the latest electronic gadget, you are ultimately contributing money to facilitate the end result of human/computer interfaces.

Competition among the consumer and commercial/industrial complex cannot remain static, i.e., without new and improved products. In order to remain profitable they have to keep upgrading and innovating. This is part of what drives Moore's Law. Moore's Law automatically renders obsolete the products that were the "hot items" just last year. It is a never-ending cycle that feeds on itself. Every time you or a company purchases the newest "hot items" in order to be more productive, part of that purchase price goes into research and development.

Included in this research and development funding are contributions from the world's governments. Some of that research and development is ultimately directed toward the development of human-enhancing cybernetics (i.e., cyborgs). This is all rooted and grounded in the advancement of information technology. Like it or not, we all have a helping hand in funding the development of these technologies.

There *will* be a physical limit beyond which we cannot go. As stated before, that limit in computer technology is estimated to be reached around 2020, provided the human race doesn't do itself in, or some other event occurs to place humans back into the dark ages … or worse.

However, one event that would put an end to all of these technologies, as they would no longer be needed, is the physical return of Jesus the Christ to planet Earth—possibly very soon.

CHAPTER 6

NANOTECHNOLOGY

WHAT IS NANOTECHNOLOGY, and how important is it? A *nano* is one billionth of a meter. A meter is 39.36 inches. This may be hard to visualize, so let's give it a different perspective. An object one-billionth of a meter in size is like comparing a marble to the Earth.

Further advancement in information technology is impossible without the development of nanotechnology. Further advancement in synthetic and artificial life is impossible without the development of nanotechnology. Advancements in genetic engineering could not have happened without the developments that have already occurred in nanotechnology. Nanotechnology has already played major roles in the development of new products in the field of chemistry. Let us not forget that all living cells are made up of chemicals and it is the development of new chemicals made possible through nanotechnology that synthetic/artificial life scientists are using as they try to create life in a test tube. Indeed, all of the sub-related fields mentioned so far in this book *cannot* continue in their advancement without the parallel advancement of nanotechnology. That is how important it is.

Nanotechnology is obviously an emerging science. With increasing ability, it can precisely manipulate matter on previously impossible scales. There are very few areas of human technology that nanotechnology does not benefit. Nanotechnology is capable of manufacturing microscopic motors, driving microscopic gears, to be used in a variety of applications. The current impact of nanotechnology and the future impact it will have on our lives is extensive and imminent. Following is a list showing the broad range of applications nanotechnology has already benefitted. This list includes (but is not limited to) chemistry, medicine, information and communication, energy, consumer goods, and heavy industry.

In the field of chemistry nanotechnology used in the form of nanoparticles can strongly influence or even change properties of material, like stiffness, elasticity, melting points, heat tolerance, or cold tolerance. Nanoparticles are revolutionizing the entire chemical spectrum with applications previously unheard of. Chemical engineers, using nanotechnology, can develop and provide tailor-made enzymes, polymers, molecules, nanoparticles, and a host of other exotic chemicals found only in living tissues. Other chemicals are completely new to the chemical spectrum. Many of these tailor-made chemicals are being used by genetic engineers.

Nanotechnology is used in filtration devices, filtering for specific contaminants, ranging anywhere from water treatment plants to kidney dialysis machines. This technology can also be used in air filtration systems. The list of uses seems endless.

In the field of medicine, the list of uses seems endless as well. There are applications of delivering drugs to specific cells, lowering the amount of drugs needed and at the same time reducing side effects. Nanotechnology has applications in physical therapy, diagnostic devices, and analytical tools. Human tissue can be regrown on carbon nanotube scaffolds. Advancements in medical research, such as in the fight against cancer, are at the forefront.

In the field of information and communication, nanotechnology is the key to this electronic engineering. Without going into great detail, nanotechnology is helping to develop quantum computers, using quantum algorithms. To help this new technology succeed there must be new developments in semiconductor devices, memory storage, optoelectric devices, integrated circuits, and a multitude of other circuit-enhancing devices. These advances will not be possible without the use of nanotechnology.

Currently, it looks as though that when the engineers get to the atomic and subatomic levels of computer technology, further advancements in this field are not in view. As stated before, the best guess of the engineers involved is that they believe this level of technology will be reached somewhere in the 2020 time frame. At this time there is no other technology on the horizon to allow the field of computer science/engineering to advance any further beyond that point.

Energy is another area where nanotechnology will play a vital role. Nano-technology will have its most important role in reducing energy consumption by increasing the efficiency by which energy is consumed. We can use this technology to make better thermal insulation. Reducing the amount and weight of materials needed in all types of construction and manufacturing will result in energy savings. As an example, light bulbs will go from five percent efficiency of turning electrical energy into light to more than 95 percent efficient. Nanotechnology is helping us create more efficient batteries for hybrid automobiles. So many applications (too numerous to mention here) will be involved in the reduction of energy, without any loss in human efficiency.

Nanotechnology will also be instrumental in increasing our food production. In part, food production increases will be because of genetic engineering, including higher crop yields across the entire spectrum of consumable plant life, along with more efficient weight gains in the livestock industry. Other consumer-oriented goods include scratch-resistant optics, easy-to-clean surfaces of ceramics, cosmetics, more efficient sunscreens, and better cleaning products. As previously stated, there will be very few areas of human technology that nanotechnology will not benefit.

Finally, nanotechnology will find magnitudes of uses in the heavy industry sector. Automobile manufacturers will make use of the lighter and stronger materials, resulting in better fuel economies and longer-lasting machines. Nanoparticles can make steel stronger, less rigid, and better able to withstand the elements—all-around improvement of steel's properties. Lighter and stronger materials will affect the aerospace industry and space exploration. Paint with self-healing capabilities and corrosion protection is being developed.

As you can see from these examples, nanotechnology will impact almost every facet of human existence. Again, as stated earlier, further advancement in information technology, genetic engineering, synthetic and artificial life, chemistry, medicine, energy, consumer goods, heavy industry, and all of the related subfields in science and technology will be impossible without the development and application of nanotechnology.

These first six chapters point out where these different technologies are converging, *if left unchecked*. At the beginning of the book, it was pointed out that these overlapping fields of scientific study are enhancing and advancing each individual branch of scientific endeavor. They are *converging*! What will the end result be?

Except for genetic engineering, which can have an immediate effect on Earth's species, most of the topics discussed thus far are really in the not-too-distant future. The converging issues that will be discussed in the next chapters have a much more imminent and foreboding outcome, should they get out of control and actually come to pass. Remember as you read the next chapters that it only takes one person in the right position to spoil everything for everybody and bring the whole world crashing down. The right person in the right position (like Adolf Hitler), armed with the technological information and know-how thus far presented, *could* wreak havoc on the whole world.

A small group of people could also use what follows in the next few chapters to take control and govern the entire world with an iron fist. The upcoming information is even more frightening to contemplate.

Are you up for it?

CHAPTER 7

TERRORISM, POLITICAL CHAOS, WAR

ALL OF THE previous fields of scientific endeavor mentioned so far are converging to a point where even though humans may benefit from enhancements in these fields in the short term, they are leading to a place we may not *want* to go in the long term. Nonetheless, they *are* converging.

There are also other converging issues that have the ability to end life as we know it on planet Earth. Unless God intervenes, the outlook for humanity is not good when these converging issues are considered. We will now begin to examine these other important scenarios and how they are converging.

The previous six chapters and their estimated timetable depend on relatively stable economic, political, religious, and peaceful environments. If any or all of these factors are in turmoil or collapse, then advancements in research and development can be impacted to the degree that such turmoil or collapse exists. Consideration as to whether these events are local, regional, national, or worldwide must be taken into account.

Let us look at terrorism, political chaos, and war. How do each of these relate to convergence?

Terrorism

What is terrorism and what does it look like? Some would say you'd certainly know it when you see it. Why would people, or groups of people, resort to terrorism? What purpose do they want to achieve? What do they hope to gain? Although definitions vary, one is that terrorism is the use of violence to create an atmosphere of fear in a targeted population, in order to bring about changes in specific political, governmental, religious, economic, or cultural objectives (sometimes all of the above).

Many times when mass demonstrations do not result in compromise or resolution of grievances, some within those groups will resort to violence in order to effect change. That is a form of terrorism. Often the violence of terrorism is perpetrated by the very few who feel they have nothing to lose and everything to gain.

For a Muslim, the ultimate gain is their heaven; for others, maybe just a better standard of living or political or religious justice. All terrorism has the intended purpose of destruction of property and the injury (or death) of innocent bystanders, no matter the reason. To a terrorist's mind, the end justifies the means and therefore any cost or harm the "means" perpetrates on the innocent. If innocent people get in the way or are in the wrong place at the wrong time, it doesn't matter to the terrorists. It is all part of the cost of doing "business." And scared populations will usually give up just about anything, or *everything,* in order to have safety and security.

Terrorism is the outward expression of anger. Look at all of the terrorism that is going on around the world today and you'll see a lot of angry people. One thing about anger, no matter the issue or cause, is that it is contagious. It can spread like wildfire.

This type of terrorism is also perpetrated by one nation against another. Cross-border excursions by armed militants in civilian dress may evoke a military response from the nation against which the terrorist attack was committed. It is a low-grade form of war. These types of low-grade wars can often lead to actual war between two nations, regional groups of nations ... or all-out global war.

Examples of terrorism abound in history past and in recent history. Just consider recent news from around the world. In order for terrorism to work, it has to cause political chaos. Political chaos can bring down governments or change governments. It can bring down dictators and it can create dictators. Many times the end results are *not* the ideals which started the movement in the first place.

In Russia during the time of the last czar, there was a sizable gap between the rich and the poor in that nation. Under Vladimir Lenin, the Bolsheviks stirred up support among the poor people by pointing out this unfair political and economic condition. Lenin promised that if the peasants would help him overthrow the czar and the government, every peasant would receive 40 acres that would be his to farm and pass on to the next generation.

The uprising was successful and the Soviet Union was born. But after a couple of years, the government (Lenin, the dictator) took the 40 acres back from everyone. All farmland was made "collective," meaning the farmers were now tenants, or sharecroppers, for the government. What was the end result? Communism. Except for Lenin's cronies, everybody was poor. There were no longer any rich people in Russia, the poor people were still poor, and everybody lost what little freedom they had. The poor people did not get what they were promised. They had just been "useful idiots."

Unfortunately, it is not difficult to see the beginning parallels of this starting in the United States as well. The seeds of discontent are being sown between the "haves" and the "have-nots." The rich are the reason for the less fortunate ... or so the accusation goes. Some call this "class warfare." People should be careful what they wish for (and riot for), because just like in Russia, what they get may not be what they wanted or were promised. Everybody could end up poor. History can and does repeat itself!

Now, we must also consider "cyber terrorism." Before we get into this type of terrorism a distinction needs to be made between "hackers," cybercrimes, cyber terrorism, and cyber warfare. *Cyber hackers* are often a teenager or young college student who, just "for the fun of it," tries to break into computer networks. They may make changes to web sites that are supposed to be funny in their minds and cause minor nuisance problems.

The next step above cyber hackers are those who perpetrate *cybercrimes*. These are the ones who with criminal intent use deception to try to get unsuspecting computer users to divulge personal information, so the hackers can steal and unlawfully use that personal information for their own gain.

Cyber terrorism is a premeditated cyber attack, mostly politically- or economically-motivated and directed against large computer network systems. Cyber terrorists like to target banking systems, power grids, and other types of utilities. These targets include anything from railroads to nuclear reactors. They especially like to target old, vulnerable computer systems of business, government, and industry. This type of terrorism is often sponsored by foreign governments seeking to find where vulnerabilities may exist. The United States government is involved in this as well. Shutting down a power grid of an enemy nation or severely limiting their ability to communicate would be very handy in time of war.

Cyber warfare attacks are often directed toward military targets (though not exclusively so). In 2011 it was speculated that first the Israelis, and then the U.S., desiring to slow down and postpone Iran's development of nuclear weapons, conducted a cyber attack on Iran's nuclear facilities. Whether they did or not, it is clear that *somebody* did, as the Iranians themselves acknowledged their nuclear computer systems were temporarily made inoperable due to an implanted virus. According to nuclear experts, a three-month delay in development was the result. *The reason the world's societies are so vulnerable to severe disruption of their everyday life is because we have become almost solely reliant on computer systems.*

Computers control our automobiles. General Motors (GM) boasts that with their "OnStar" system, if a car is stolen and it is reported to OnStar, they can then cause the automobile to stop running and also lock the doors, preventing the thieves from escaping! Pretty cool, right? But suppose somebody had a grudge against GM, for whatever reason, hacked into their On Star computer system, and caused every GM-manufactured vehicle with On Star capabilities to stop running? Can you imagine the traffic gridlock that would cause across the entire country? Would that not work for terrorists or an enemy country, to cause a national emergency?

Computers control our banking system (which includes credit card transactions), the Wall Street stock trading system, our military communications and linkages, and our phone company computer networks (which give us access to the Internet). Computers control our electrical power grid and so much more. What do they all have in common? They all run on electricity. Would a rogue nation like Iran, China, Russia, or North Korea want to see our power grid shut down, given the right conditions (in their minds)? If America's nation-wide power grid was made inoperative for just one week, it would bring the country to a standstill immediately, with the likelihood of total breakdown of civil obedience.

Think for a minute what a catastrophe that would cause … there would be no gasoline pumped. No banks open for business. No factories operating. Hospitals could operate only until their diesel-powered generators ran out of fuel. The perishable foods at the grocery and in your refrigerators and freezers would spoil. There would be no computer systems—with all they control—operating anywhere. There would be no running water in towns and cities and no well water in the country, where pumps run on electricity. Humans can't live much more than a few days without water, let alone the water needs of livestock. Can you visualize the mass hysteria, looting, and killing in the name of survival?

I can hear some people now, saying, "That can't happen. There are too many security fire walls in place to keep such things from happening. Computers systems are constantly being upgraded and new software put in place so that the unthinkable can't happen." Really? Even so, no matter how tight the security, no matter how current the computer, nothing is 100 percent safe. There are always "counter" measures to the measures.

Even the Germans in World War II thought their "enigma" coding and decoding machine was foolproof. Yet, unbeknownst to the Germans, early on the British broke the code. It is no different today with so-called computer "security" systems. Our enemies out there right now may have already broken the computer security codes and passwords, and be waiting for the right time to break in and disrupt ... and we wouldn't have a clue that this had already happened.

"Stealth" is just as important to cyber warfare for undetected entry into enemy cyberspace as it is for stealth aircraft to be invisible to radar, with the purpose of undetected entry into enemy airspace. Many nations have the supercomputer technologies needed to accomplish the disabling abilities of code-breaking and virus-planting. It's been reported that a Chinese cyber warfare squad took over and brought safely to Earth one of the United States' high-tech stealth drones that was flying over Iran. Do you think the U.S. military was shocked to realize that this could (and did) happen?

Political Chaos

Political chaos is brought about by large groups of extremely unhappy and angry people. Many times people are angry and they don't even know why, or what their anger is about. They just want "change." The mind-set is whatever the change, it has to be better than what they have now. Mob demonstrations are usually what occur first. When the violence occurs with crowd push-back from possibly over-zealous police or soldiers (and violence is often one of the goals of mob demonstrations), then bystanders sympathetic to the demonstrations many times join in. At this point mass demonstrations accompanied by civil disobedience often occur. The mass demonstrators, right or wrong, always feel they have just causes.

All of this "churn," this political unrest, not only causes political chaos, but it can also lead to civil war and then to the overthrow of democracies, capitalism, communism, socialist governments, and dictators. The recent history of the so-called "Arab Spring" is a good example of terrorism, mob violence, and mass demonstrations working together to effect major change in national goals and directions. When this kind of disenchantment among populations becomes widespread across the world, unintended consequences could prove to have further consequences, triggering events that lead to out-of-control scenarios that then lead to war ... even world war.

War

War can be local, regional, or global. Unfortunately, war is as old as mankind. The specific causes vary. A cause could be border disputes and national boundaries, or it could be differences in political ideologies between nations. War has the potential to spread among and between nations due to political and military alliances. Sometimes it is due to greed for power, accumulating territory, or wealth. The egos of kings and dictators and heads of state get involved. Sometimes war starts over long-held grudges, with attempts to settle "old scores" that may go back hundreds of years. The militant expansion of religious beliefs from religions that believe in "evangelism with the sword" has caused many wars. Sometimes, all of the above have a role.

One thing about war is that technology has enabled the combatants to progressively kill people in ever-increasing numbers through genocide or on the battlefield. The ability to leave cities in total ruins was demonstrated in World War II. Towns and cities were destroyed from England to France to Germany to Poland to Russia to Japan.

The world is currently on a "knife edge," and major war could break out at any time. Too many scenarios are converging in that direction. The winds of war are increasing. Major conflicts may have already broken out by the time you read this. Once the "war genie is out of the bottle," it is very difficult to get it back in.

In the late 1960s I had the opportunity to visit Nagasaki, where the second nuclear weapon from the United States was detonated. It is estimated that 70,000 people were instantly vaporized. There are displays of shadows of people on whitewashed walls and shadows of footprints on the concrete sidewalks, where people were standing or walking when the bomb went off. Schoolchildren in Japan wore uniforms much like what the Navy wore. They were either all dark blue with white trim, or they were all white with blue trim. There was a group of schoolchildren waiting to cross a street when the bomb detonated. They were far enough away from ground zero that they were not incinerated. However, they were close enough for the intensity of the light and heat to burn them. The dark uniforms absorbed the intense light and heat and burned the children so badly most did not survive. Conversely, those children wearing the mainly white uniforms received severe burns where the dark trim was, but the mainly white parts reflected enough of the heat and light energy that they mostly survived.

The nuclear weapon that was detonated over Nagasaki was just a fire cracker in comparison to the nuclear weapons that are in abundance around the world today. There will be more on this subject in a later chapter.

Future Wars

There are future wars prophesied in Holy Scripture that will be more terrible than anything mankind has experienced in all of recorded history. First, we will look at Ezekiel 38 and 39, Psalm 83, and Revelation 9:15–16.

Ezekiel speaks about a future war known as the "Gog-Magog" war. (This war is not to be confused with Armageddon.) The Gog-Magog war will occur most likely before (or right at the beginning) of the tribulation period, also foretold in Scripture. Ezekiel names the coalition of nations that will be involved in this war. Of course he used their ancient names, but for this study their modern names will be used for the sake of clarity.

The countries named in Ezekiel are Russia (possibly meaning just southern Russia), Iran (which was Persia up until 1935), Sudan, Ethiopia, Libya, and possibly Algeria and Tunisia (all African nations), and Turkey and Armenia, meaning people of Central Asia. Psalm 83, speaking of the same war, adds Jordan, Lebanon, Syria, and Saudi Arabia. Notably missing from this list are Iraq and Egypt.

For the first time in history, this coalition of nations listed in Ezekiel and Psalm 83 is actually coming together (at the time of this writing) and forming military alliances—*for the express purpose of destroying Israel,* "that the name of Israel will be remembered no more"(Psalm 83:4).

In order for all of this to happen, Israel has to exist as a country, located in the Land of Israel. For 2,000 years, the Land of Israel lay vacant and desolate. However, just as God promised many times in the Old Testament that He *would* bring them back to His Land from the far corners of the Earth, in 1948 Israel once again became a nation. Read and hear what God says through Ezekiel in 38:14–16:

> Therefore, son of man, prophesy and say to Gog: This is what the Sovereign LORD says: "In that day, when My people Israel are living in safety, will you not take notice of it? You will come from your place in the *far north,* you and the many nations with you, all of them riding on horses, a great horde, a mighty army. You will advance against My people Israel like a cloud that covers the land. In days to come, O Gog, I will bring you against My land, so that the nations may know Me when I show Myself fully through you before their eyes" (emphasis added).

Isaiah 17:1 speaks specifically of Damascus, the capital of Syria, as does Jeremiah 49:23. These verses say that Damascus will no longer be a city, but will become a heap of ruins. Even the walls of stone will be set on fire. This has not historically happened. This prophecy will more than likely be fulfilled during the Gog-Magog war, or possibly during the war of Armageddon.

The whole Book of Revelation chronicles the events leading up to Armageddon. Revelation 9:15 says that "at that time," there will be 200 million troops poised to cross the Euphrates River when it dries up. *This war will involve all nations of the Earth.* Jesus Himself says in Matthew 24:21–22, "For then there will be great distress, unequaled from the beginning of the world until now—and never to be equaled again. If those days had not been cut short, no one would survive, but for the sake of the elect those days will be shortened."

Jesus is telling us that mankind, left to its own devices, will eliminate all traces of humanity from Earth through these atrocious wars. *If God does not intervene, that will be mankind's destiny. It is what we are converging towards.* These wars will be more terrible than anything mankind has experienced in all of history and we now have the weapons and the technology to do it. *If there is no God, then mankind has no future hope of anything.* But Jesus says that He and His Father *will* intervene and cut that war short.

Remember this statement: *"God is ultimately our only hope."*

CHAPTER 8

WEAPONS OF MASS EXTINCTION

Nuclear Weapons

WHEN WE HEAR the phrase "weapons of mass destruction," most people think of the nuclear bomb. Indeed we should. The previous chapter gave an example of the destructive power of nuclear weapons. Since the atom was first harnessed in 1945 and used against the Japanese empire, many nations have since acquired the technology and ability to produce atomic weapons. Besides the United States, these nations include Great Britain, France, Russia, China, India, and Pakistan. More recently, we can also add the nations of North Korea and (since the breakup of the Soviet Union) Ukraine, Belarus, and Kazakhstan. Iran seems to be the latest to acquire the materials, knowledge, and weapons. There is evidence that Russia, China, and North Korea have shared their technological knowledge with Iran and others with the money to pay for it. Some of the fissile material was unaccounted for after the breakup of the Soviet Union and was feared to have ended up with black marketers and rogue nations. This is in part what gave rise to the term "nuclear proliferation."

Several of these nations are known to be nations of Islam. As radical Muslims, they are committed to helping each other destroy the nation of Israel "that the name of Israel be remembered no more." (Refer to Psalm 83:1–5.) Their stated goals are not only to destroy Israel, but also to destroy Israel's most ardent supporter, the United States. There are several ways they might attempt this.

Ever since September 11, 2001, the United States has been quite concerned with nuclear weapons "slipping into" the country. Government and security officials are concerned about container ships entering the ports or terrorists coming across the Mexican or Canadian borders, and any other way that they could see that a rogue nation (or terrorist group sponsored by a rogue nation) could slip a suitcase-size nuclear weapon into the U.S. for the purpose of mass destruction and murder. Under such a scenario, millions of people could die.

Homeland Security, the FBI, and the CIA, as well as other state and local government officials, are all keeping a lookout for so-called "sleeper cells." Nuclear fissile material the size of an orange could wipe out one square mile of densely-populated downtowns of major cities. Setting off a nuclear weapon is the same as creating a miniature man-made star. It is the same nuclear reaction that takes place on our sun. Such an event would be instantaneous and without warning.

At one time, the former Soviet Union had between 300 and 1,500 nuclear weapons targeting American cities, capable of total destruction of those cities. That would be every major and medium-sized city in the United States and then some. The nuclear fallout would take care of the rest of us. Those weapons still exist and can be pointed at the U.S. in a matter of minutes. This could happen by accident. Nuclear annihilation could occur by miscalculation of adversaries, or by sheer madness.

There has been at least two times, once in the 1980s and once in the 1990s, when Russian technology indicated that the U.S. was launching nuclear missiles at them. On one occasion, Russian President Boris Yeltsin had his special suitcase open. *All he had to do was push the button to release a retaliatory strike.* Thank God he could not bring himself to actually push that button! It turned out to be a false alarm—but that is how close we came to an "accidental war" with nuclear weapons.

While I was in Japan many years ago, I read a book written by a Japanese doctor (*Hiroshima Diary: The Journal of a Japanese Physician August 6-September 30, 1945,* Michiko Hachiya, M.D.), who had survived the nuclear blast at Hiroshima and had treated many of the survivors. The book was written from his day-to-day journals and included pictures. Following is a short account of some of his observations.

The Japanese doctor stated that many individuals who were relatively close to ground zero, but not so close as to be incinerated, had burns and broken bones. Most eventually got better. They did not display symptoms of atomic radiation. He described many who were not injured by the blast, but began showing symptoms of radiation sickness, which he did not understand at the time. He described, and the pictures confirm, that people of all ages began having nausea and diarrhea. This began two to three days after the blast. Next, as the symptoms persisted, small areas of blood spots developed under the skin. From there, the skin began to deteriorate and rot, until it would hang from all portions of the body and the person would die. It was a slow and very painful death. These effects were not confined to people alone; animals and vegetation also suffered from the effects of atomic radiation.

It doesn't sound logical that those who were closest to the blast did not come down with radiation sickness, while those who were much further away did. However, the explanation is simple. The blast threw tons of radioactive material straight up into the atmosphere and debris then rained down in the shape of an umbrella—or mushroom—coating the outlying areas and people with radioactive debris.

These are the real results of just one "small" nuclear detonation.

Then, there is the 40- to 50-mile radius around the Russian nuclear power plant in Chernobyl; that meltdown occurred in 1986. This area has been, is now, and will be for quite some time a "no man's land." Some of the people suffered immediate symptoms and death from radiation sickness, while others did not display any symptoms until many years later, based on their proximity in distance to the reactor. Twenty-five-plus years later, people can go in for short periods of time, but staying overnight is not something an intelligent person would want to do.

In March 2011, a huge magnitude 9.0 earthquake followed by a devastating tsunami heavily damaged several Japanese nuclear reactors. The full results of these meltdowns are not known at the time of this writing. However, tuna caught off the coast of California have been found to be radioactive—though so far not in amounts harmful to humans.

Several nations have nuclear weapons that can kill by radiation alone, but leave the buildings and other hard assets intact. There are prophecies in the Bible that give details of a future war where the results sound eerily similar. For example, Ezekiel 29:9–12 speaks of Egypt becoming desolate. We say "becoming" because the quote that you are about to read has not happened in all of history to Egypt:

> Egypt will become a desolate wasteland. Then they will know that I am the LORD. Because you said, "the Nile is mine; I made it," therefore I am against you and against your streams, and I will make the land of Egypt a ruin and a desolate waste from Migdol to Aswan, as far as the border of Cush [Ethiopia]. No foot of man or animal will pass through it; no one will live there for forty years. I will make the land of Egypt desolate among devastated lands, and her cities will lie desolate forty years among ruined cities. And I will disperse the Egyptians among the nations and scatter them through the countries.

At the time Ezekiel wrote this, the name "Aswan" was not in existence. You can look on ancient maps. It is not there. From 1958–1970, the Russians helped the Egyptians dam up the Nile River. It is the largest dam in the world and it is called the Aswan High Dam. This dam has created one of the largest man-made lakes ever. On either end of the dam are two large towers. "Migdol" is the Hebrew word for "tower." The lake that this dam formed is called Lake Nasser. It is 340 miles long and 22 miles wide.

The Aswan High Dam is so large and well-constructed, it is said, that in a time of war, *only a large nuclear weapon could destroy it*. With the way things are going in the Middle East it is certainly conceivable that this could happen. What would be the result if that dam was suddenly destroyed? With 340 miles of lake behind it, everything within twenty miles of the shores of the Nile River from the dam to the Mediterranean Sea would be wiped out, just from the water alone. Cities would be destroyed and the land would be made desolate. About 80 percent of Egypt's population lives within that corridor. Then there would be the radioactive pollution in the water and the atmosphere. What the water didn't destroy, the radiation would make useless for many years.

This may *not* be the way the cities of Egypt are destroyed and laid waste. Only God knows! But we do know that this hasn't happened yet in history, and it *is* conceivable that it could happen this way. The end result would be the same as Ezekiel described.

Zechariah 14 speaks of the last war leading up to the second coming of Jesus Christ. The symptoms described in verses 12, 13, and 15 seem very similar to the symptoms experienced in Hiroshima and Nagasaki:

> This is the plague with which the LORD will strike all the nations that fought against Jerusalem: Their flesh

will rot while they are still standing on their feet, their eyes will rot in their sockets, and their tongues will rot in their mouths. On that day men will be stricken by the LORD with great panic. Each man will seize the hand of another and will attack each other. A similar plague will strike the horses and mules, the camels and donkeys, and all the animals in those camps.

This *is* the world war against Israel that Jesus Himself will cut short, before all flesh is consumed. Jesus said that "no flesh would survive if these days were not cut short." *Remember: we have the technology to cause these things to happen and nuclear proliferation is just one aspect of this technology.*

There is another way nuclear weapons could be used. If a powerful nuclear device was exploded high in the atmosphere over the United States, there is a force given off called "EMP." This stands for Electro-Magnetic Pulse. This is an electrical pulse generated by the explosion. It would blanket the nation. This pulse would not only knock out the *entire* electrical grid, which it is designed to do (consult Chapter 7 on cyber warfare), it would also "fry" all of the electronic components in our automobiles, computers, radios, TVs, and anything else with electronic circuitry. Not many would survive in the Stone Age that would follow!

Biological Weapons

Biological weapons, as with any modern weapon for large-scale destruction, can be used locally, regionally, or globally. They can be used as a weapon of terror, as was the case following the September 11, 2001, terrorist acts, when anthrax spores were sent to various individuals through the regular mail. Biological weapons certainly can be used in times of war. Modern biological weapons have the potential to create loss of life equal to, if not greater than, the nuclear threat. Generally speaking they are easy to produce, easy to store, and easy to disperse among populations. Biological weapons are by far less expensive to produce and maintain than nuclear weapons. For this reason, many more nations possess biological weapons then nuclear weapons.

Biological weapons can be made up of viruses, bacteria, or fungus. They can be of a kind where, once a person is exposed to it, that person can then expose other people. It can spread through the population quickly, person to person.

There are other biological types that may affect a person, but cannot be transmitted to another person. Most of these contagions can be spread through the explosion of an aerosol-type device. Due to the scientific advancements in genetic engineering as mentioned in the early chapters of this book, the ability to slightly change the DNA of already known pathogens causes most vaccines and antibiotics to become useless against genetically-altered viruses, bacteria, and fungus. Humans would literally have no built-up immune system defenses to resist such genetically-modified organisms. Such would be the case with synthetic and artificial life sciences creating artificial biological warfare agents.

Besides attacking the human population, all of these biological warfare agents can also be made to attack animal, plant, and marine life.

Chemical Warfare

Chemical engineers using nanotechnology techniques can create all manner of life-ending nanoparticles capable of infecting the lungs or being absorbed by the skin, with doses no larger than the dot over the letter "i" required for a lethal dose. The number of new tailor-made agents with the ability to do this grows every day. This type of chemical agent could not be spread by contact with other human beings and thus would have a limited area of coverage. This would, nonetheless, be very useful in battlefield engagements. Even the so-called "old" chemical weapons were very useful for this purpose. Consider the Iran-Iraq war. In the 1980s, the war between these two nations was long and bitter. Iraq made extensive use of mustard gas during that war. They used it against the Iranian soldiers and also against their own people, the Kurds, with great and devastating effect.

Mustard gas was first used in World War I when the Germans used it against the British. Contact with mustard gas caused blisters on the skin. If it got into the eyes, it could cause blindness. When inhaled, the blisters would form on the inside of the tracheal tube and the lungs, causing the lungs to be ineffective in absorbing oxygen. It would also cause the lungs to fill with mucus, slowly drowning the individual. Death would occur in about 12 hours, and it was a slow and painful death. *Iraq used this 100-year-old technology.*

What is chemically available now, with the new chemical technologies, is really unknown at this point. However, we *do* know that most of the armories of the world stockpile chemical weapons.

Chemical and biological weapons, unlike most nuclear weapons, leave buildings and other assets intact. A conquering nation would only have to clean up millions of rotting corpses to inherit turnkey, user-ready buildings, equipment and other hardware.

Described in this chapter are horrible weapons of mass destruction. It is information we would rather not think about. These weapons of mass destruction have only seen very limited use, *so far*. However, we can only come to this logical conclusion: that any nation—or group of nations—when threatened with total annihilation by another nation or group of nations, will use any and all weapons at their disposal to prevent being wiped off the face of the Earth.

Given the past and current propensity of the human race to engage in war, the use of these horrific weapons is not a question of "if" but "when." Considering the increasing anger in the populations of the world, the world is converging toward destruction. Revelation 6:8: "I looked and there before me was a pale horse! Its rider was named Death, and Hades [Hell] was following close behind him. They were given power over a fourth of the earth to kill by sword, famine and plague, and by the wild beasts of the earth."

Yes—the information in these chapters is extremely depressing. It really doesn't get any better. Many of the technologies listed so far can and will have benefits for the human race. But even with thousands of people using hundreds of thousands of technologies for the benefit of mankind, it only takes one mentally-disturbed individual to bring the whole thing crashing down.

These possibilities are reality. They are a real and present danger, whether we know about them or not. *Not knowing about them or not thinking about them does not make them go away.* Sure, it "helps" us mentally and psychologically not to dwell on these things, but we are assuredly currently living in a time that is leading to what God calls "the great tribulation" (Revelation 7:14). The prophecies for the future recorded in the Bible look more and more like they *will* be fulfilled ... *and be fulfilled literally.*

God provided an "escape mechanism" for Noah and his family, and God *will* provide an escape mechanism for living believers in Jesus, at *His* appointed time. That escape mechanism is known as "the rapture" (Genesis 7:7, 1 Thessalonians 4:13–18). How much man-made war, famine, and plague we will have to endure until then is unknown.

Trust Jesus in all things! *He is our only hope.*

CHAPTER 9

THE POPULATION TIME BOMB

IN THE FALL of 2011, the Earth's population reached seven billion people. Except for the time of the Bubonic Plague, when the population receded, the human population has been steadily growing since the 1400s. Rather than going back to Adam and Eve and tracing the growth of the population from there, we will start this presentation in the 1800s. In 1804, it is estimated that the Earth reached its first one billion people. That means it took from the time of Adam and Eve until 1804, which would have been at least 6,000 years (and maybe more) to reach the one billion milestone.

Then, it only took from 1804 until 1927—123 years—for the Earth's population to reach two billion. Thirty-two years later in 1959, the population stood at three billion. The four billionth person was born 15 years later, in 1974. Next, it only took 12 years for the Earth to gain its five billionth person, in 1987. From 1987 until 1999, Earth gained another one billion people, making six billion inhabitants. Then in 2011, we reached seven billion—and counting. Another 39 years from 2011 brings us to the year 2050. The population of the Earth is estimated to be between nine and eleven billion people then. Are you beginning to see some of the possible implications of this trend?

Another way of looking at this is counting the number of years it takes to double the population of the Earth. From one to two billion, it took 123 years. Then it took 47 years to go from two billion to four billion. It is estimated the Earth will reach eight billion people by the year 2025, which means it would take 51 additional years to double. While it took 123 years for Earth to gain an additional one billion people, it was only 12 years in gaining the latest one billion.

How has this population "explosion" occurred? In the years leading up to 1804, including ancient history times before Jesus Christ, the population of the Earth was pretty well kept in check by numerous factors. Even though the natural birthrate was much higher in those days, the natural infant mortality death rate was also much higher. Many children died during birth and many mothers died giving birth. There were also many diseases going around then, with no medicines available to counteract the virulence of the diseases.

Then there were the wars. There were the Egyptians fighting the Hittites. The Assyrian Empire gained in strength and conquered surrounding countries, enslaving the survivors (which included ten of the twelve tribes of Israel). The kingdom of Babylon, under King Nebuchadnezzar, marched its armies across the Middle East and North Africa, taking many captives (including the remaining two tribes of Israel) and leaving many hundreds of thousands dead in their wake. Many kings of those ancient countries dreamed of acquiring empires for their own self-gratification. There was the Persian Empire's (which is present-day Iran) conquering of the Babylonian Empire (present-day Iraq). This was followed by Alexander the Great, the ancient Greek leader, conquering the Persian Empire and marching all the way down to Egypt. Not to be outdone, the Roman Empire came on the scene and conquered what was left of the Greek Empire.

Can you imagine the carnage that took place with all of these empires constantly on the move, conquering and killing? Whatever the numbers, those opposing soldiers and citizens never had the chance for reproduction. Many captive children, who survived the military sieges (like the prophet Daniel), were *surgically* prevented from reproducing. One must also consider the religious practices of that era, where babies and young children were actually *sacrificed* on the many altars dedicated to pagan gods. With all things considered, it is no wonder the population of the Earth took so long to reach its first one billion people. All of these factors served to keep the human population growth in check.

The human species has had very few years of total peace from the beginning of the human race. Even in recent history, with the march of Napoleon, the United States Civil War, and World War I as examples, there is proof that war has worked to prevent our population from being even greater than what it is. It is estimated that 50 million people died in World War II. Twenty million of those were Russian and six million were Jewish. It has been reported that Josef Stalin, *after* the war, killed another estimated 20 million of his own countrymen.

Since the United States Supreme Court ruled on *Roe v. Wade* in 1973, an estimated 45 million babies have been terminated. From first trimester to partial-birth, these infants were prevented from seeking life, liberty, and the pursuit of happiness, along with untold millions of abortions elsewhere around the world. In the United States, the courts routinely hold as "unconstitutional" state laws trying to ban some abortions. Unknown numbers of infants have been prevented from being conceived around the world due to birth control. China has a much-debated "one child per couple" law.

Yet *still* the Earth's population is exploding. How many people are too many? From the point of view of many in the world, the United Nations' mandates on birth control "education" (including access to free, unlimited abortion) in Third World countries are high on the agenda.

Many elected officials and others in elite positions of authority believe they can see where the population explosion is leading us. They do not like the converging issues they see and they are running scared. Try as they might to manipulate and hold back what they see coming, they will not succeed. If one does not believe in God, then mankind is ultimately left to their own devices to solve these many converging issues. Mankind does not have a good track record when it comes to peaceful solutions.

We can see what is holding back the growth of the population somewhat. But what is it that is enhancing the growth of the population? For one thing, more people surviving to reproductive age leads to the birth of more babies. Since the late 1940s when medical science came up with the first effective and reproducible antibiotic, many, many antibiotics are now on the scene preventing death from pneumonia, smallpox, measles, malaria, cholera, and a whole host of other life-threatening diseases. Clean drinking water, medical advancements in immunizations and other treatments, and chemical control of mosquitoes and other disease-bearing insects keep people living longer.

Many are worrying about rising "greenhouse emissions" caused by all these people. They see human activity as the cause of global warming/climate change. Many scientists believe that our only hope of escape, seeing the collision course that we are on with the population, is to "think" our way out of this population explosion by using technology (Transhumanism?). The scientists recognize we are on a collision course—a train wreck, if you will—because of the dwindling resources of our planet. Can we really "think" our way out of what is coming?

Isaiah 13:9–13 records:

> See, the day of the LORD is coming—a cruel day, with wrath and fierce anger—to make the land desolate and destroy the sinners within it. The stars of heaven and their constellations will not show their light. The rising sun will be darkened and the moon will not give its light. I will punish the world for its evil, the wicked for their sins. I will put an end to the arrogance of the haughty

and will humble the pride of the ruthless. I will make people scarcer than pure gold, more rare than the gold of Ophir. Therefore I will make the heavens tremble; and the earth will shake from its place at the wrath of the LORD Almighty, in the day of His burning anger.

This prophecy has not yet been fulfilled. We are converging towards something, and it doesn't look pretty. The following chapters will consider some of the "dwindling resources" that cannot keep supporting a growing population.

CHAPTER 10

OUR FRAGILE PLANET

Food Production

AS THE POPULATION of the Earth continues to double rapidly, so must the production of food also double. The food staples that the world depends on the most are wheat, corn, rice, and soybeans. Currently, the population is increasing at a pace greater than our ability to increase food production.

What are the early indications of food becoming harder to obtain for the masses? One of the earliest signs is price inflation. Like other commodities, food increases in price as supply lessens. It is said that the civil disobedience that began in Tunisia and triggered the so-called "Arab Spring" was over the people not having enough money to buy the basic food sustenance needed for their families. While food inflation in the United States is something for most every household to complain about, in Third World countries it is a matter of life and death. In the Third World, it is a matter of having the necessary monies to purchase enough food to sustain life for oneself and one's family. Money is most often in short supply for these people living where 60 to 70 percent of their income goes toward food.

In the chapter on Genetic Engineering, it was revealed that many of our food crops have been genetically-modified. These crops are known as GMO's, genetically-modified organisms. This has greatly increased our ability to produce food and to keep up with the food demands of the world. Yet, a recent survey indicated that the United States has a 35-day food supply, while the remaining world has a 15-day supply. These are the lowest numbers since these types of records have been kept—and these numbers do not bode well for the future.

What are some of the leading contributors to food shortages? As just discussed, population increases and the resultant increase in food consumption place a huge burden on mankind's ability to produce an adequate supply of food. There are many other causes of food shortages, such as floods. Even just abnormal amounts of rain, causing the ground to be too wet, can prevent farmers from having the ability to get into fields to plant crops. The other extreme is not enough rain, where the crops shrivel and die due to lack of moisture. Extended periods of drought where rainfall used to be the norm can also cause regional food shortages. Underground aquifers used to irrigate crops are being depleted. Weather patterns seem to be changing. Extreme weather in the United States has had a steep impact on food production that is now seen in increased prices at the supermarket. We have to also consider that we are "burning up" our food: we are converting corn and other food staples into ethanol, in order to supplement oil production, which is having a difficult time keeping up with the growing demand.

Then there is war. Food shortages can lead to war and invading armies can destroy food crops in the field and in storage. It is a way to demoralize and defeat your enemy and has been used as a weapon throughout history.

Finally, what would *seem* to be a minor issue is in reality having quite an impact: that is the disappearing populations (up to a 50 percent reduction) of the honeybee, worldwide. Scientists are not sure why this phenomenon is happening. The evidence is pointing toward a fly parasite, or larva, invading the abdomen of honeybees and eventually killing them. Honeybees help crops to pollinate. Without pollination, most crops cannot produce food. At this time there is *not* an artificial substitute to take the honeybees' place.

As time goes on, our food shortages can only get worse. There are many implications in this for our species. If people in the United States are so willing to stampede, trample, punch, gouge, and fight while "waiting in line" to buy a limited amount of expensive tennis shoes, or participate in the "Black Friday shopping experience" as "hockey player wannabes," running over everybody else to be the first to get in the store, *how will they act when food shortages are real*?

Oil Reserves

Believe it or not, oil production also impacts the food supply. The development of the internal combustion engine, and how to put it to use for mankind, was a direct result of the discovery of oil. The internal combustion engine was an exponential technological advancement over the horse and plow. It allowed one man to produce food for not only his family, but also enabled him to produce food ultimately for hundreds of families and in much less time. Oil-based pesticides contributed to the increase in food production. Much of the fertilizer used around the world is oil- or gas-based. Anhydrous ammonia is used in great quantities and has been one of the key factors in the increase of food production. It is a very powerful and potent fertilizer that can change poor soil into super soil for one growing season, and is essential to higher-yielding crops. Anhydrous ammonia is made from hydrogen. Hydrogen is made by breaking down the water molecule, H_2O. This process takes either natural gas via steam reforming, or electrolysis using electricity. Either way, the energy that is used to create the hydrogen, and thus the anhydrous ammonia, is far greater than the amount of energy in the hydrogen that is produced.

Hydrogen would be a great fuel to power our internal combustion engines in cars, trucks, and other transportation modes where reliable and renewable sources of energy are needed. But since it takes more energy to produce liquid hydrogen than what it gives off, it is a "dead-end" as a replacement for oil. As oil reserves decline, the price will increase ... and so will the cost of food. *They are converging.*

Human population growth has placed the greatest demand on oil. The human population continues to grow at the same time that oil production has peaked. Known oil reserves are in decline and will not keep up with the demand in a very few years. New, unknown oil finds *may* delay the inevitable for a few years. However, the end scenario remains the same. This scenario will bring about not only unemployment, but unprecedented famine in the decades to come. Nations will fight over what oil and food remains. We are *not* talking about *arguing*—we are talking about *war*. Do you perhaps think that may be a little overboard? Consider the following.

In 1956 France and England went to war with Egypt because Egypt closed down the Suez Canal and stopped the flow of oil to those countries. The war didn't last long, but it was over oil. Obviously there were other underlying issues that precipitated this conflict, but for Egypt to cut off the oil supply was a stranglehold that could severely cripple and shut down the economy of most European nations, even back then. The ability to respond militarily would be severely limited, with only hours' to a few days' oil supply in reserve. This was most definitely a threat to the national security of France and England; cutting off their oil supply was perceived as an act of war by these countries.

In 1940, the United States was not happy with Japan for invading China. The Japanese received a major portion of their oil from the United States. In order to help persuade the Japanese to withdraw from China, President Franklin D. Roosevelt implied that the U.S. *might* start reducing the amount of oil shipped to their country. Japan interpreted that as a serious threat to their national security ... and Pearl Harbor was the result, drawing the U.S. into World War II.

The United States currently finds itself dependent for *60 percent* of its oil supplies from foreign countries.

Thirty-five percent of that comes from the Middle East. If oil coming from the Middle East were suddenly to come to a halt, what would that do for U.S. food production, and its ability to produce and transport any and all types of goods and services? Much of the food in our supermarkets travels an average of 1,000 miles—mostly by trucks, which use diesel fuel. Do you think cutting off one-third of the oil supply would endanger national security? Do you not think the U.S. would go to war in the name of national survival?

Obviously there are sources of alternative energies, but can they (or will they) come online in time and in the amounts needed to fill the demand vacuum that the lack of oil would leave behind? I personally do not think so. The U.S. can certainly go into a conservation mode. One way governments can enforce conservation is to drive up the cost of oil and gasoline by using taxes and regulation, forcing many to drive fuel-efficient autos and use energy-efficient lighting or other energy-saving systems (all at much higher costs for consumers), but that would only help to postpone the inevitable.

Oil affects almost every aspect of the economy. It affects not only food production, but essentially *every aspect of our lives*, in one way or another. We live in an oil-based economy. High energy prices slow down economic activity, causing unemployment.

The following statements might be considered radical. These things would not happen overnight; they are simply intended to demonstrate our dependence on oil. Consider: When oil stops, the economy stops. When oil stops, electricity stops. When electricity stops, water stops. When oil stops, food stops.

Look at it this way: The Earth's exponential population growth coincided with the discovery and development of oil. Without oil, the population could not have been supported and growth would have been severely limited. What happens to the population when oil is depleted? It doesn't take a rocket scientist to see how this is all *converging*.

Oil is not a renewable resource. When it is gone, it is gone ... and so is a major portion of the food supply. Nuclear energy cannot fill the gap. Solar energy cannot fill the gap. Biofuels will by necessity have to stop. Wind energy cannot fill the gap. Synthetic fuels cannot fill the gap. All of these alternative fuels taken together *cannot* fill the gap, once our non-renewable oil source is depleted.

Population growth is not the only thing putting extra demand on all types of non-renewable fossil fuels. Consider just one country: India. If every one of the one billion people in that country wanted to increase their electrical consumption by just a single 100-watt light bulb, that is equivalent to 100,000,000,000 watts. This would take an additional 200, 500-megawatt nuclear or fossil fuel-fired generating power plants just to meet the additional demand. Extrapolate this to the population of all Third World countries and you can see how quickly non-renewable resources can be used up.

New technologies such as fusion (which is the opposite of nuclear fission) are not holding out much promise. The greatest promise in this area is the fusion process using helium. The problem with this is that the only source for this helium, in quantities large enough to produce the kind of power the Earth would need, is only found in the rocks on the moon.

Back in the 1980s a scientist announced he had developed a way to produce hydrogen gas by using a technique known as cold fusion. However, this could not be duplicated by other laboratories around the world. Had this process worked, it would have been a boon to our energy future. Hydrogen gas would have produced much more energy than what it took to produce the hydrogen gas.

Other Resources?

The resources of the Earth come in two forms. They are either *renewable* or *non-renewable*. Examples of renewable resources would be wood and paper products from forests. By replanting the forest with new trees, these kinds of wood product resources can have extended lifetime benefits for human consumption. Currently, on the non-renewable resource side, there is starting to be a shortage of copper. Other metallic resources, such as aluminum and iron ore, are becoming more expensive and harder to process from the Earth's crust. Most of the easy stuff has already been mined, leaving the more difficult to process. That's why we have been in a recycling mode on many of these non-renewable resources. Increased usage of these categories of resources will ultimately deplete known sources. There *is* a limit beyond which the Earth *cannot* maintain those renewable and non-renewable resources needed to satisfy the demands of its growing population.

Jesus tells us in Matthew 24:21 and 25, "For then there will be great distress, unequaled from the beginning of the world until now, and never will be equaled again. See, I have told you ahead of time."

Jesus said there are terrible times coming. We are in a position to see the *convergence* of so many Earth-shaking events! Trauma and severe human suffering are poised to shake us like we have never been shaken before. There will be huge reductions in the Earth's population. Time estimates vary. It could easily be within your lifetime. *It is not a matter of "if," but "when," these things come to pass.*

You may say, "Life is good! Mankind has always found ways to solve these dilemmas." But hear this reasonable warning: *only God will be able to solve this last dilemma.* If you've been a skeptic maybe this has helped to snap you back into reality. Life is good ... but for how long? There are basically only two avenues that humans will take when the time comes, when reality is at their front door, ready to invade their lives. It will be either: "Eat, drink, and be merry, for tomorrow we die," or else people will turn to the Almighty God in a serious way. Many will ask Jesus to come into their hearts. Most will not. *Which will it be for you?*

CHAPTER 11

THE ANGRY EARTH

THERE ARE PEOPLE who make a living looking at trends, then making projections into the future as to where these trends will eventually end up. These people are known as *Futurists*. Most are involved in *micro* projections. Micro projections are myopic in nature. Their view is short-term and narrow in scope. They look at various product trends, advising corporations about where to emphasize their energies. They identify which products will be successful, the number of potential sales, and which products, even though technically attractive, will ultimately flop.

In writing this book, I am doing the same thing, but am using *macro* projections. Macro projections look at the big picture. I am projecting where these macro trends are heading and what their ultimate conclusion will be. This process requires looking at not only the current conditions and issues, but requires going back in history—sometimes ancient history—to project the long-term *or* short-term outcome. During this process, as many sources as possible should be consulted. There are trends that are so obvious that it doesn't take a rocket scientist to see where things are going.

You have no doubt realized by now that I am not a "fluff" writer. I stick pretty close to the facts and to projections made by reliable sources. Everything covered so far is easily verifiable. Obviously, there are differing opinions and conclusions. I am attempting to not be an optimist *or* a pessimist, but am rather trying to allow the evidence and the trends to speak for themselves.

I am also pointing out that many of these trends, and their *converging* conclusions, *were predicted thousands of years ago*. The ancient biblical prophecies are looking more and more as though they will be literally fulfilled. As these prophecies concerning the Earth and mankind are shown to be accurate, should we not also consider and take seriously other truths contained in the Holy Scriptures, truths that have previously been dismissed by man? Fact: the Bible is the only "Holy Book" that makes such predictions.

You might not accept the Bible as being credible. You might rather accept how some interpret the Mayan calendar or the allegorical interpretation of Nostradamus' quatrains, making them say whatever the interpreter sees as authoritative for future predictions. However, it is hard to deny the uncanny similarities in what the Bible literally describes with what can be seen emerging and converging at this moment in history.

It *is* Going by the Book

The Bible gives signs and conditions—*not* a timeline—although many have tried to make it give a timeline in the past (setting dates) and will no doubt continue to do so in the future.

The Bible does speak about conditions on the Earth that will take place leading up to and during end times. We should not easily dismiss such a resource. One of the things the Bible speaks of is the increasing frequency and severity of natural disasters. In Luke 21:10–11, 25–28, and 35, Jesus speaks of the signs concerning the end of the age:

> Then He said to them: "Nation will rise against nation, and kingdom against kingdom. There will be great earthquakes, famines and pestilences in various places and fearful events and great signs from heaven.... There will be signs in the sun, moon and stars. On the earth, nations will be in anguish and perplexity at the roaring and tossing of the sea. Men will faint from terror, apprehensive of what is coming on the world, for the heavenly bodies will be shaken. At that time they will see the Son of Man coming in a cloud with power and great glory. When these things begin to take place, stand up and lift up your heads because your redemption is drawing near." ... "For it will come upon all those who live on the face of the whole earth."

That is plain English: great earthquakes, great famines, great pestilences ... biological war? It does not have to be allegorized to make it say something else. Are the natural (and possibly unnatural) disasters listed here in Luke in fact happening in the recent past, and even as we speak? Let us examine the facts.

On the surface, one could make an argument that we have always, from the beginning of time, had earthquakes and floods, wildfires and tsunamis. We have always had droughts and famines, volcanoes and tornadoes. This certainly is true. But we want to know if this is starting to happen on a larger, world-wide scale, with increasing (that's the key word ... "increasing" ...) frequency and severity.

Severe Weather Events

Let's start with severe weather events and their related effects. In the year 2010 alone, 263 million people were affected world-wide in one way or another by weather-related disasters. A small percentage, which still amounts to tens of thousands of people, were killed. The rest were either injured, made homeless, forced to live in tents and other temporary shelters, displaced, or evacuated. Many suffered from lack of food, lack of clean drinking water, and lack of adequate clothing and blankets. The physical destruction caused by any natural disaster sometimes takes years to rebuild. The emotional scars and effects can last a lifetime.

The year 2011 was called "the year of the tornado" in the United States. One website estimates that 840 tornadoes, with 552 people killed and an estimated $27.3 billion in property damage (a record-setting amount) occurred in that year.

Weather-related disasters can be caused by tornadoes, floods, droughts, windstorms, lightning, hail, blizzards, hurricanes, and wildfires. Speaking of windstorms, lightning, and hail, Revelation 16:21 speaks of a time when a thunderstorm will be so severe that "from the sky huge hailstones of about a hundred pounds each fell upon men." Every year since 2000 has seen record-setting weather-related events somewhere in the world. Weather-related victims are expected to increase to 375 million by the year 2015. Some of this increase can be attributed to increasing population, some to increasing severe weather events.

These numbers of people affected by "weather-related" disasters do not include those casualties from earthquakes or volcanoes or tsunamis. When we do include volcanoes, earthquakes, and tsunamis, the death toll and the destruction have increased dramatically in the last decade. Consider Indonesia, Japan, and Haiti ... to name a few.

Some say global warming/climate change is the reason for the increase and severity of severe weather events. Many of the scientists and self-appointed Earth protectors have used the term "global warming" for many years to scare people into changing their energy consumption behavior. For various reasons, the term "global warming" lost its traction and credibility, so a new term had to be developed. The new catchphrase is "climate change."

It may very well be that the Earth *is* warming at an accelerated pace. After all, the Earth has been warming ever since the last ice age. It may very well be that human activity is contributing somewhat to this phenomenon. The jury is still out when it comes to all scientists believing that global warming is solely the result of human activity, since there has been no increase in average global temperatures in the past ten years. Those who do believe it are pointing to the burning of fossil fuels as the culprit. Whether we stop burning fossil fuels or we run out, the outcome for the population of the world is the same ... and it isn't pretty. The Bible also points to a time when the Earth will be suffering from scorching heat, in Revelation 16:8–9: "The fourth angel poured out his bowl on the sun, and the sun was given power to scorch people with fire. They were seared by the intense heat and they cursed the Name of God ... but they refused to repent and glorify Him."

All life on Earth is obviously dependent upon the sun. Our lives depend on the warmth of the sun's rays. Both the National Academy of Sciences and NASA have issued statements concerning the probability of a major "solar flare" event. They warn that it would cause trillions of dollars in damage by knocking out the world's power grids and take up to ten years to recover from such an event. They even go so far as setting a timeline of May 2013, which coincides with Solar Max, with the peak of the sun's eleven-year cycle of solar flare activity occurring in 2013/14. Can we count this as one of the *converging* events?

Whether man-made or naturally-occurring, severe weather has been on the increase. Even record cold snaps were caused by global warming ... or so they say. But what about earthquakes, tsunamis, volcanoes, and the dreaded asteroids? It has been reported that natural disasters of all types around the Earth have increased 300% in the last thirty years. It can't all be blamed on global warming.

Earthquakes

Some websites contend that there is no *significant* increase in earthquakes in the last 150 years or so. They then go on to show 38-year segments, beginning in 1863, of the numbers of recorded earthquakes of magnitude 7.0 or greater:

    1863–1900   12 earthquakes
    1900–1938   53 earthquakes
    1938–1976   71 earthquakes
    1976–2011   123 earthquakes—with estimates of 180 by 2014.

I don't know about you, but that seems like a significant increase to me.

Some in the scientific community want to attribute these increases to better detection systems. While that could be true in the early years, beginning in the late 1940s the United States seismic detection network was sensitive enough to pick up nuclear test explosions both above ground and below ground, a half-world away in the Soviet Union. Thus, from the 1940s to the present, the definite increase in 7.0 or greater earthquakes (which are far more Earth-shaking than nuclear explosions), cannot be attributed to more sensitive equipment.

In the year 2000, 1,505 earthquakes of magnitude 5.0 or greater were recorded. In the year 2010, 2,427 of 5.0 or greater were recorded. That's an increase of 922 earthquakes ... in just one decade. Is that not a significant increase? One of the most severe earthquakes ever recorded hit Japan in March of 2011. Besides the initial damage from the earthquake, the anomaly of the huge tsunami that followed created even more devastation. The death toll ranged upwards of 20,000, with the destruction of thousands upon thousands of homes and also a nuclear power plant occurring. The economy of Japan and the reduced output of its factories created far-reaching shortages even in the United States.

We can safely say that earthquakes are on the increase in severity and frequency. The Bible tells us about a massive, global earthquake in Revelation 16:18–19, referring to events that will occur during a seven-year period called the tribulation: "Then there came flashes of lightning, rumblings, peals of thunder and a severe earthquake. No earthquake like it has ever occurred since man has been on the earth, so tremendous was the quake. The great city split into three parts, and the cities of the nations collapsed." Talk about the ultimate in severity! An earthquake of that magnitude could easily create a 300-foot high wall of water, going inland for who knows how far. We already know what destructive, lethal tsunamis with walls of water 10, 15, and 20 feet high can accomplish. In a moment, we will speak of another possible event that could indeed cause a 300-foot-high wall of water to be created.

Volcanoes

Although the Bible does not speak of volcanic eruptions being on the increase, they are related to the movement of the tectonic plates. The movement of these ground masses certainly causes most earthquakes. As earthquakes have increased recently, so have volcanic eruptions.

Even volcanoes that have been asleep for hundreds (if not thousands) of years are awakening, with Washington State's Mt. St. Helen's in 1980 and Mt. Rainier in 1991 as examples.

Wyoming's Yellowstone National Park has been determined to be one large volcanic cauldron. The volcanic magma is so close to the surface there that steam vents—including Old Faithful—bubbling, sulfuric mud pits, hot springs, and such are the norm. The floor of Yellowstone Lake has been rising and getting warmer, indicators of the pressure buildup occurring in that area. It could theoretically blow at any time, but scientists *think* it *could* be decades, or maybe even centuries away. When it does go off (and it surely will), it would throw so much debris into the atmosphere, blocking the sun's rays, that scientists say it would cause another Ice Age. We are told that Yellowstone has erupted three times in the prehistoric past.

Two hundred years ago volcanoes around the Earth erupted about 17 times per year. Looking at a ten-year segmented graph, every ten-year period since then has shown increases in volcanic activity. The total of volcanic eruptions around the Earth currently stands at about 70 per year.

The Dreaded Asteroids

Until the last couple hundred years, humans did not know what an asteroid was. Today, with our highly-advanced methods of detecting asteroids using sophisticated radar and other detection methods, we have come to find out the Earth is not always alone in its orbit. We know from looking at the craters on the moon that asteroids have been out there and active in past history. We have also seen the evidence of impacts here on the Earth: there is Crater Lake in Oregon and a mile-wide crater in Arizona. Scientists and geologists tell us that a giant asteroid struck the Yucatan Peninsula of Mexico 65 million years ago. That asteroid was larger than Mount Everest and estimated to be six miles across when it hit. It is credited with wiping out the dinosaur population.

From 1997 to the year 2000, there were zero detected asteroids in near-Earth orbit. In 2001 there was one, 2002 two, 2003 five, 2004 eight, 2005 six, and 2006 twelve. As of the writing of this book, there are more than 4,000 asteroids—and counting—in the near-Earth orbit. The average size is 100 feet across. If it struck the Earth, a 100-foot asteroid could easily wipe out a medium-size city. Scientists are worried that such an event could occur and are developing plans to use rockets, nuclear weapons, or other space-bound technologies to prevent an asteroid from striking the Earth. That is how seriously they are taking the threat.

When the Bible was being written, mankind didn't have "asteroid" in their vocabulary. They had no idea such a thing existed. Listen to what God revealed to the apostle John in Revelation 8:8–9: "The second angel sounded his trumpet, and something like a huge mountain, all ablaze, was thrown into the sea. A third of the sea turned into blood, a third of the living creatures in the sea died, and a third of the ships were destroyed." This sounds somewhat like what hit the Yucatan Peninsula, don't you think?

Can you imagine the atmospheric shock wave such an event would cause? How high of a wall of water would be generated from such a large object crashing into the ocean at a horrific speed? This is what was revealed to John concerning the end time tribulation period. John didn't know what an asteroid was. He only knew it was the size of a mountain. We know that it's a future asteroid, because John said it was all ablaze. Atmospheric resistance would cause such a phenomenon. Nothing else will fit the description. At the time of this appointed encounter, it is obvious that mankind will not have the technology available to push this giant asteroid off course. For this reason, *it must be going to happen relatively soon.*

If mankind knew enough about weather patterns, if we knew how the sun's rays and its energy react and interact with all the facets of climatology, if we could precisely take all of the necessary climatology readings and know precisely how cause and effect produces an outcome when it comes to our weather, then we could precisely predict the weather indefinitely into the future, even to 1,000 years or more. Sadly, we do not possess the ability or the technology to even get a five-day forecast right.

Consider, however, that there *is* an all-knowing, "other-dimensional" Intelligence, communicating these and other events thousands of years ago, with 100 percent accuracy, to selected individuals. This "other-dimensional" Intelligence has revealed ahead of time all aspects of the progression of Earth's activities, culminating and *converging* in what we are beginning to see transpire. This is absolute evidence of His existence.

Or ... do you perhaps think it is all just coincidence? If you do, you should consider the statistical improbability—no, *impossibility*—of that opinion.

CHAPTER 12

IT'S THE ECONOMY, STUPID ...

THIS PHRASE WAS first coined when George H. W. Bush was the 41st President of the United States. Since the inception of this country, our economy has had cycles of boom and bust. There were times when the economy was roaring and there were times of great depression. One seemed to always follow the other. Everybody likes it when the economy is good; people are employed and all is well with the world. But economic pullbacks, the recessions in the economy, especially severe recessions, can lead to depression. That causes concern and alarm for most people.

During those times of high unemployment, when people are losing their homes and jobs, is when "disgruntled" people look to vote for politicians who promise the most—including the quickest economic recovery. Currently, politicians who promise the longest periods of unemployment benefits and the most welfare benefits are gaining more and more support at the voting booths. At the time of this writing, approximately 49.5 percent of Americans pay *no* income tax. Most of those receive benefits from the government. What happens to the country if this voting bloc becomes the majority, voting only for candidates who will increase their benefits? Some people work for a living; some vote for a living. There is much government can do to alleviate some economic problems ... and there is much the government can do to exasperate or add to the problems.

What influences the direction of the economy and can we ultimately control the up-and-down cycles, control the volatility, keeping it all at a more acceptable level? Over the decades, educated economists certainly have put forth their theories and policies, trying to keep the economic engine running smoothly. So far, none have worked as promised.

This is not intended to be an in-depth study of economics; it is only to highlight how some economic trends can be manipulated for better or for worse. Ultimately, in a free society, the economy goes in the same direction the masses of people believe it will go. If the population's view is that money is going to "get tight," that the economy is going to slow down, then people will start preparing for just such an outcome. They will put off buying that new car or house or appliance, all because they believe something is going to stall the economy. It is somewhat like a self-fulfilling prediction. People work to pay down debt, instead of buying new things and taking on more debt. If enough people believe this way, the economy will most definitely slow down. This type of economic pullback is the result of the masses believing that an economic pullback is imminent. A pullback of this nature is probably not based on economic reality. Once the momentum starts in either direction—up or down, it is hard to contain.

In a government-controlled economy, the people have little say in economic direction. This kind of control may sound inviting for those tired of the economic "roller coaster ride." It is attractive in the short-term, but has been shown to fail, as in the collapse of the Soviet Union. China had to allow its economy to become more free market-based in order for the government to ultimately survive.

"It's the Economy, Stupid" is the title of this chapter. George H. W. Bush was President in 1991 when the United States went to war with Iraq. This war was known as the Gulf War. At the end of the war, President Bush had a very high approval rating among voters. The economy was doing very well. The President was up for re-election and looked to be a "shoo-in" for a second term.

So, how do you defeat a popular, first-term president? Political pundits know that presidents are elected or turned out of office based on economic realities. While the troops were still coming home, the national media started talking about how the economy was "pulling back" ... when in reality it wasn't. Night after night, for months, they kept hammering the idea that the economy was going downhill quickly. Guess what? People started to believe it. They started preparing for it out of fear of not being able to "weather the storm." By the end of the year, the economy was looking bleak. The President and the Republican establishment were pointing to all of the good accomplishments of the President. However, they were not addressing the economy, which by then was in the uppermost parts of most peoples' minds. Hence the phrase ... "It's the economy, stupid."

The manipulation of the economy *by the news media*, coupled with third-party candidate Ross Perot, caused a once-popular President to lose a second term ... all in the name of politics and power ... but at whose expense? I would suggest at the then-newly-unemployed peoples' expense. Bill Clinton was elected President and proved to be *somewhat* capable in terms of the economy; however, similar circumstances in the early 1930s in Germany brought Adolf Hitler to power. Be careful who you elect in a bad economy. It could be the undoing of a democratic, capitalist nation ...

Obviously there are other ways the economy can be manipulated, mostly by the government and also, in America's case, the Federal Reserve. How? Adjusting interest rates up and down is an attempt to keep the economy growing without either hyperinflation or deflation. Regarding the money supply, actual printing of money, along with creating money out of "thin air" (like telling a government computer to deposit "x" number of dollars in someone's account, when the money didn't exist to start with), is *creating* money. The United States government has bailed out banks and large corporations with the help of this technique.

This works because people still have confidence in a piece of paper that costs five cents to produce, no matter the denomination, and that the government *says* has value ... *when ultimately it has none.* Paper money used to have value when it was backed by gold and silver, but now they say it is backed by the so-called economic prowess of the United States. In reality, the paper dollar will only have value as long as the masses believe it has value.

One thing is for sure—the more money we print and create, the less value it will have and hyperinflation can be expected. When money loses its value, it takes more of it to buy the same thing that could be bought for much less in earlier times. This is okay if wages keep up, but it kills those who are savers and those who are on fixed incomes, like retirees. The government finds it easier to pay off its debt with inflated dollars at the expense and punishment of savers and retired people. A person used to be able to buy a new car for what it now costs to buy a riding lawnmower. That's inflation!

To hedge against inflation, people are into purchasing precious metals such as gold and silver. However, there is a time coming on the Earth where, according to what God revealed to Ezekiel, "They will throw their silver into the streets, and their gold will be an unclean thing. Their silver and gold will not be able to save them in the day of the LORD's wrath" (Ezekiel 7:19). Revelation 6:6 reveals the ultimate in inflation: "Then I heard what sounded like a voice among the four living creatures, saying, 'A quart of wheat for a day's wages, and three quarts of barley for a day's wages, and do not damage the oil and the wine!'"

We have a debt-based world economy. When we borrow, we are using money that we have not earned yet. We are borrowing against the future: future jobs as individuals and a future, taxpaying generation for a nation.

When a nation, any nation, goes into debt to a greater extent than it can tax the people or print money, that nation will default on its debt, causing an economic collapse. Once a nation borrows more than it can repay it has only two options, both leading to collapse. It can either continue to borrow and print more money, or it can cut back on its governmental services and try to repay the debt. Once you borrow beyond a certain point, you can only postpone the inevitable by borrowing.

Perhaps I can explain this in another way. I learned to fly when I was a young man. When I was taking multi-engine instruction, one of the great aeronautical principles was drummed into my head by the instructor. When you are coming in for a landing, you have the flaps extended and the landing gear down. This makes the airplane what they call "dirty," in that there is a lot of induced drag that the airplane's engines have to overcome. It is critical not to let the aircraft slow down below a certain speed. Bad things happen if you do! Besides the obvious, which is an instant crash, there is a phenomenon known as "getting behind the power train." This is where if you slow beyond a certain point, even though the aircraft continues in controlled flight, adding full power to the engines will not arrest the descending of the aircraft. If the aircraft is too low on approach to the runway and you're going too slow, you are going to impact the ground before you get there. Nothing you can do will change that outcome. Talk about ruining your day! This is sort of the way it is when a nation—or an individual—borrows more than they can repay. Like the twin-engine airplane, you can go down quickly or less quickly, *but go down you will.*

Recently, I ran into a lady who was trying to decide if she should spend her last dollar and 39 cents on a cup of coffee. She turned to me and said "I am on the verge of filing for bankruptcy. What's another dollar and 39 cents?" She bought the coffee.

World economics are tied together. They are globally interconnected. When one or two countries fail economically, that affects *all* of the national economies of the world. It is contagious! *We have the beginnings of a one-world economy.* This can be easily seen by watching the stock markets of the world. When the United States market takes a sudden dip, so will the Asian and European markets. When their markets take a dip first, American markets will follow the next trading day. It has been stated that when the United States catches an economic cold, the whole world comes down with the flu.

World economic chaos could soon be upon us. According to Scripture, there is a time coming (known as the "tribulation period") when the economies of the world will, by necessity, be centrally-controlled. They will be controlled by one man, a powerful dictator, who will "force every one, small and great, rich and poor, free and slave, to receive a mark on his right hand or on his forehead, so that no one could buy or sell except he had the mark ..." (Revelation 13:16–17). *Mankind currently has the technology to allow this type of economic control to happen.*

As demonstrated in earlier chapters, severe weather events, earthquakes, tsunamis, a sudden dip in world oil supplies, and war (limited or all-out) can all have a devastating effect on the economies of the world, outside of the normal economic schisms. These things can slow the economy or even bring it to a virtual standstill. This kind of economic disaster, along with other *converging* items occurring simultaneously, can and *will* have dire, unforeseen consequences.

In the preface of this book, I gave my definition of *convergence*. I present it again: "Individual objectives moving independently toward a point of focus or outcome; progress or technical achievement becoming more frequent and more intense as time goes on; and distance to the outcome becoming shorter. The individual lines of convergence will gain momentum (reference Moore's Law), if nothing occurs to slow it down."

World events point toward an extraordinary chain of catastrophes moving toward us at an accelerated pace. Every person, in their gut, knows we are on the *converging* verge of war, food shortages, oil shortages, and genetic manipulation out of control in the form of killer viruses and genetically-engineered chimeras. The weather is going crazy, earthquakes are increasing, there is population explosion, nuclear proliferation, worldwide economic and social turmoil, rejection of biblical authority by much of the Church ... and *immorality* is now the new norm for *morality*. The hatred of Israel by *all* nations (and especially Islamic nations) is and will be a flash point. And there is so much more.

What is the world going to do? What can *we* do to stop the direction the world is going in? How can we be saved from ourselves? There are so many apocalyptic problems converging on the horizon in the near future. Which ones will detonate the events that lead to the apocalyptic end?

A college friend of mine who is very much into video games told me that almost all violent video games have one person/character come to the rescue to save the day and the human race from total annihilation. Oh, how we need such a person!

We need a powerful, super-intelligence to shake us into reality, someone who can set us back on the right track, to let us know we are not alone in the universe. We are lost as a human race, looking at the very distinct possibility of extinction. We need to know that there is a higher, more intelligent life form that has successfully overcome these problems we face in the universe.

We need a savior ... oh, how we need a savior!

CHAPTER 13

UFOs TO THE RESCUE!

OH, HOW WE need someone, anyone, to take us back from the abyss, to infuse us with new and better technology! To show us how to solve our energy problems, that the human species might evolve and take our rightful place in the cosmos. Is that too much to ask?

Modern man has been searching the cosmos for a long time, trying to detect life forms in our own universe, as well as farther out into the cosmos. The search for extraterrestrial intelligence (SETI) has been going on around the globe since the early 1960s. All this is an attempt to discover "intelligent life" beyond the bounds of Earth. Our governments and scientific institutions are spending billions of dollars looking for this "extraterrestrial" life. Even the Vatican has announced that the idea of "ET's" living on other planets is highly probable.

With all the research that governments and the scientific community are conducting, at the same time they seem to be *denying* the possibility of alien UFOs (unidentified flying objects). But many credible witnesses to the UFO phenomenon are coming forward with their experiences. The general consensus is that they are real, they are from other intelligent civilizations, and they have the means and technology to possibly cross thousands of light years in distance to explore planet Earth. However, we don't know who they are, what they are, or where they come from. Are they friend or foe? What are their intentions? Are they coming to save us or enslave us? Are they coming to serve us, observe us ... or to conquer us? It has been speculated that if they meant us harm, it would have happened already. Of course, they could have discovered that we just aren't the tasty treat they were expecting!

In order for intelligently-operated alien UFO craft to physically exist, to be who (or what) most think they are, there has to be some assumptions made. The first assumption is that there *is* life elsewhere in the universe, where that life has evolved to a high level of intelligence. This is an overwhelmingly accepted theory of belief. The second assumption is that they have found ways to overcome many of the laws of physics that seem to be barriers to us.

What have the observers told us about the attributes of these UFOs? To begin with, they are visible and they come in a variety of sizes and shapes. They have variously been described as disk-, saucer-, or cigar-shaped, with lighted windows; large triangular crafts, spheres, cylinders, walnut-shaped, upside-down Reese's Cup, egg-shaped, and teardrop-shaped. They range in size from several feet to larger than several aircraft carriers combined. Another reported attribute of UFOs is their astronomical speed, where they have gone from zero miles per hour to out of sight in a second or two.

Besides extreme rates of speed, they have extreme rates of climb and extreme rates of maneuverability that are completely unmatchable by any aircraft known to exist on this planet. The witnesses have reported that they heard no wind sound from the craft going through the atmosphere at high rates of speed, and no propulsion noise. People in 1800s Europe were reporting "mystery airships" with these same characteristics, including observed battles between craft.

Who are these modern-day credible witnesses? They are 747 captains, Air Force fighter jet pilots, policemen, high-ranking military officials, a state governor, and even former president Ronald Reagan ... and so many more. Their eyewitness accounts cannot be taken lightly.

On average, there are about 500 "UFO" sightings per month around the world, most of which can be *rationally* explained. There have been numerous documentaries on various cable channels, TV shows, movies, books, and Internet stories portraying UFOs. The documentaries all seem to highlight and re-enact UFO encounters with humans. There are many organizations, governmental and private, dedicated to investigating UFO reports. One such report, from the UK's Project Condign, concluded that several aircraft have been destroyed over Russia, with at least four pilots killed chasing UFOs. Did you ever hear about that? Many of the visual reports were confirmed by radar returns, along with aircraft intercepts.

There are three more so-called "incidents" we will briefly highlight, to use as examples for the purposes of this chapter. All of these incidents and many others as well can easily be verified on the Internet.

The "Rendlesham Forest Incident" occurred in December 1980, just outside the perimeter of the U.S. Air Force Base in England. UFOs were reported on two different nights by military police and the deputy base commander, United States Air Force Lieutenant Colonel Charles I. Halt. Lieutenant Colonel Halt made a real-time tape-recording of his conversations and conversations of others, as they were viewing and describing the close-up encounter with a UFO. Colonel Holt has signed an affidavit, verifying what he called the "extraterrestrial event" of that evening. He also gave a personal testimony of that event in 2007 at the National Press Club in Washington, D.C.

The second example is known as the Belgian UFO wave. It occurred from December 1989 until April 1990. On one occasion in particular, there were eyewitness reports of UFOs in the area. This apparently was confirmed by simultaneous radar returns. Two F-16s were dispatched to intercept and investigate the UFOs. According to the pilots' testimonies, they in fact did see the unidentified crafts, gave chase, and even obtained a radar lock to fire missiles at the unidentified flying objects. The pilots reported that as soon as they obtained radar lock, that lock was immediately broken due to the erratic maneuvering of the target. The UFOs accelerated out of sight within a second or two. It has been said that when a radar lock occurs, for the opposing aircraft it is a death sentence. The lock cannot be broken. Yet these UFOs broke the radar lock.

The last example is known as the "Phoenix lights incident." Apparently thousands of people viewed this UFO incident over Phoenix on March 13, 1997. They saw a vast triangular, v-shaped object gliding slowly and silently across the Phoenix sky. Witnesses estimated that the object was larger than many football fields; some estimates were up to a mile in length.

Former Arizona Governor Fife Symington III admitted ten years later that he also saw the unidentified flying object. His description, recounted in a TV documentary on the incident, was that "it was enormous and inexplicable." Governor Symington moderated the Washington Press Club gathering of November 14, 2007, where mostly retired military men and pilots from around the world gathered to give their UFO experience and testimony. Governor Symington said his UFO experience was dramatic. He stated the UFO had a geometric outline and a constant shape. Other witnesses reported that it was clearly a solid, technological flying machine that blocked out the visibility of stars as it silently passed overhead.

Certainly these four examples have their naysayers who try to explain away all UFO phenomenon that were witnessed. Yet, we clearly cannot dismiss the credibility of these eyewitness reports. What in the world ... or *out of this world* ... is going on?

None of these particular examples recount any personal contact with alien beings. However, hundreds (maybe even thousands) of first-hand reports do recount what is known as "alien abductions." While none of these reports will be highlighted here, we will note that there seems to be a common thread among the stories related in these cases. Most were taken aboard a craft and physically examined (mostly in a sexual way) and biopsies taken. They all said they were mentally alert but physically paralyzed. Almost all expressed great fear or anger. These are just some of the common experiences. There are others. Does this sound like "friendly encounters"?

In other cases, however, though not taken aboard a craft, some have reported being in very close proximity to a craft, with a feeling of great peace and assurance of no harm. Thus, while there seems to be mostly negative encounters with UFOs, there evidently have been *some* positive encounters.

It was mentioned earlier that the maneuverability of these so-called UFOs was far beyond any capability of any craft known to man. They could go in one direction at thousands of miles per hour, instantly stop and go in another direction at the same speed. If we humans would try to perform maneuvers like this while operating a flying craft of some type, the gravitational forces (which would be astronomically high) would cause us to be a thin film on the inside wall of that craft. Whatever these beings are, they are apparently able to overcome extreme gravitational forces.

The United Nations has begun to take these UFO reports, along with the discovery of other Earth-like planets in the cosmos, much more seriously. Astrophysicist Mazlan Othman, who in 2010 was the head of the United Nations Office for Outer Space Affairs (UNOOSA), was then subsequently appointed as the first UN official responsible for representing humanity *in the case of contact with extraterrestrial life*. Ms. Othman told fellow scientists in September 2010 that mankind needs to be ready to deal with alien contact.

Since 1947, there has been an ever-increasing flood of UFO incidents, UFO documentaries, science fiction movies, and science fiction TV shows. Are we humans *longing* for extraterrestrial contact? At the same time, are we being prepared (or perhaps *programmed* is a better word ...) for just such an event? Are we being desensitized to not only "long" for such contact, but to expect it? *Are we being deceived*, so that this UFO phenomenon can gain an accepting foothold when it does occur? It certainly seems, with the proliferation of all this "UFO mania" that is flooding our media outlet markets, that we humans *are* being set up. *But for what purpose?*

Suppose for a minute that one of these alien UFO craft actually landed on the White House lawn. Having neutralized all of the White House defensive measures, "humanoid" figures emerge from this craft and make themselves understood in our language. What if they told us that *they* created all life on Earth; that *they* were the ones who seeded life on this planet? That everything we humans are, we owe to them? *Is that not the role we have always ascribed to God?*

What if these aliens told us that the extraterrestrials were the ones who genetically manipulated us, and that they came to give us peace, prosperity, and knowledge? What if they told us that they came at this "critical crossroads in our history," to help us on our evolutionary course, to help humankind take its place among the civilizations of the cosmos?

What would we do? What would we think? What would *your* reaction be? Would you believe them? Would this at least *appear* to confirm to people that life on this planet did not happen by chance, by accident, or by random chemicals coming together to form the first cell of life? What would that do to the Darwinian Theory? Would that not trash the Darwinian Theory? Would not all of the religions of the world have to re-assess and re-evaluate the claims of those religions? For many Christians, their belief that God created the heavens and the earth, with mankind as His crowning achievement on Earth, would likely be re-evaluated.

But consider: *What if all of this was just plain deception, to make the world believe in one thing, when something else is actually true?*

In the documentary film *Expelled* featuring Ben Stein, Mr. Stein had an interview with Richard Dawkins, the renowned physicist of our time. In this interview, Mr. Stein, pointing to the evidence that life on this planet could *not* have started by chance happening, that there was so much design incorporated into that first cell of life, asked Mr. Dawkins if he believed that a supernatural power, even God, was responsible for creating life on this planet. Mr. Dawkins vehemently denied any such possibility. To him and many others, God just does not exist. Such a consideration was not in his realm of thought.

However, when Mr. Stein asked Mr. Dawkins if an "extraterrestrial" from a highly-advanced civilization on another planet could have been responsible for the finely-tuned design of life on this planet, that perhaps *they* are the ones who started that first spark of life on this planet and have routinely visited this planet to genetically direct our future, Mr. Dawkins quickly and very enthusiastically agreed to this possibility.

Mr. Dawkins knows that chance happening can't explain the complex design and variety of life proliferating on the Earth. There *had* to be a highly-intelligent designer at work behind the scenes.

Talk about intellectual hypocrisy! It is just plain unscientific to *not* consider all possibilities. With this admission, Mr. Dawkins is apparently willing to scrap much of the Darwinian Theory. So many scientific intellectuals will vehemently defend Darwin's Theory of evolution, *until* some "extraterrestrial" comes along saying otherwise! Very interestingly, Stephen Hawking, another renowned physicist with similar views to Mr. Dawkins, has advocated extreme caution in regard to trying to make contact with so-called "extraterrestrial life." He fears they might not be friendly!

Early in this chapter it was explained that, in order for intelligently-operated alien UFO craft to physically exist, there has to be an assumption that there *is* life elsewhere in the universe, where that life has evolved to a high level of intelligence. Please refer to Chapter 4, where there is a discussion about laboratories trying to create life in a test tube ... that even with the known chemicals and the sterile laboratory environment, with the manipulation of these variables by highly-intelligent human beings, they still have yet to achieve a single self-reproducing living cell from scratch.

The chances of random chemicals coming together in the contaminated primordial soup of planet Earth or any other planet that may be out there, no matter how many, no matter if there are millions of billions of them, are so incredibly remote that it is still mathematically impossible for life to have occurred here—or anywhere—by chance.

If it is *mathematically impossible* for life to have happened by chance, *then these extraterrestrials and their UFO craft are not from other galaxies or solar systems*. If they are *not* from other galaxies or solar systems, then *where are they from, what are they up to* and *why the great deception*? We humans seem to have bought into this "alien worldview," this belief system, this *alien deception* that is flooding upon us like a great tsunami. *Again, if they are not extraterrestrial ... then what are they*?

I would submit to you that they are NOT extraterrestrial, but *"extra-dimensional." They are from another dimension.* They have the ability to navigate between our dimension and theirs.

Let us consider some of the attributes these crafts could have, if they were from another dimension. Let us explore how overcoming our laws of physics is a non-issue with them, while those laws are major stumbling blocks to us.

First of all, by being from another dimension, they are not subject to our laws of physics. They may, no doubt, have physical limitations in their dimension, but these do not carry over or apply to our dimension. For example, they would not be subject to our law of gravity. This could explain a lot of maneuvers that mankind has observed these UFOs performing. UFOs have been seen floating effortlessly above the ground, lacking any noise of propulsion, and, in some cases, so huge as to defy description.

For us, the laws of physics are just that, laws. They are immutable for us as humans. Obviously there are more laws of physics involved in this UFO issue than what is highlighted in this chapter. What is presented are just a few of the obvious ones. An engineer I once worked with was trying to convince me that there are no "absolutes." I suggested to him that he should go up on top of the building and jump ... then come back and tell me that there are no absolutes. To begin with, even his statement is hypocritical. His statement alone reveals that there is at least one absolute: to say that there are no absolutes *is an absolute!*

There *are* certain absolutes in our physical universe—*constants*, meaning that they do not change ... *constants* that dictate the ways we can and cannot operate in our environment/dimension.

Here is another law or absolute. A body at rest tends to stay at rest; a body in motion tends to stay in motion. Imagine for a moment you are going down a ski slope. You're traveling maybe 30, 35, 40 miles per hour. You lose control, go off the course and slam head-on into a tree. This has happened to many over the years and the outcome was not good; in some cases, it was even deadly. Earlier in this chapter it was noted that if we were in a craft that went several thousand miles per hour, then immediately stopped and went in another direction at the same (or greater) speed, that we would be no more than a thin film on the inside of that craft. It would be like the skier running into the tree, but multiplied to the "nth" power. Our bodies could not withstand such a violent maneuver. We would be reduced to the most basic building block of our body, the atom. This tells us that the alien being from another dimension may have a body, but it is *not* physical like ours, does not have mass as we know it and therefore is not subject to the laws of motion.

If the makeup of the alien beings and their craft is not physical in nature, then gliding silently through our atmosphere or traversing the skies at high speed would not present a problem for them. No sound would be produced because there would be no interaction between our physical air and their non-physical craft. They would need no noisy power system to overcome atmospheric drag, because there would not be any drag.

By not being subject to our laws of physics, they would not be subject to the speed of light, which has been measured at 186,000 miles per second. That is the fastest known entity in our dimension/cosmos. In their dimension, there may not be any speed limit. This means that they could traverse our dimension/cosmos at will, all at speeds that outpace the speed of light by magnitudes of power. How fast does the speed of "thought" travel? I can go any distance forward or backward in time in an instant, at the speed of thought. This is much faster than the speed of light. Might this also apply to the extra-dimensional beings and their craft? Our thoughts are not subject to our laws of physics … just a thought. The extra-dimensionals also seem to have the ability to appear and disappear at will.

Our physical dimension has boundaries. There are attributes that define our physical dimension. Height, width, length, and time, define and encompass our entire cosmos—all 14 billion light-years of it. If alien craft are from another dimension, they may not have any of these barriers at all. If these beings and their craft are not subject to these boundaries in their dimension, they certainly will not be subject to them in ours. They obviously have a different makeup than we do—a different body mass, operating within a different reality.

Time is the last element in our dimension we will discuss in this chapter. Time can also be a factor of speed. Albert Einstein theorized that time and the measurement of time was affected by speed; i.e., the faster you go the slower time proceeds. He theorized that if we could travel at or near the speed of light, then time would practically stand still for the traveler.

If there were a set of twins where one remained on earth and the other did a round trip of one hour at or near the speed of light, the one who took the trip would have aged one hour, while his twin remaining on Earth would have aged 40 years. Therefore, for us, time is relative and not necessarily a constant.

What does this have to do with our UFO traveler? Well, since he's from a different dimension, there may be no such thing as time, or at least time as *we* understand it. For them, our time is irrelevant. Time is something that was created at the creation of our cosmos and is part of our physical laws. What was there before time began? Did the other dimension exist then? If it did (and I believe that it did) and there was no concept of time there, then we could safely say it had no beginning and will have no ending.

Concerning the alien abduction cases, I mentioned common threads earlier. These cases are what I call "catch and release," and generally leave behind negative—even fearful—feelings and memories, when recalled under hypnosis. Whether they were abduction cases or just close encounters, people cannot account for what seems to be missing periods of time in their lives. Are their encounters so close to the other dimension that for them time ceases? If this other dimension has no concept of time, no concept of a beginning or an ending, could we say that these UFO travelers are coming from what would seem to us as eternity?

This "other dimension" scenario answers many of the questions of how alien UFOs are able to do the things they are reported to do. It allows us to better understand how they can overcome our laws of physics and how the alien occupants of a craft can keep from being a thin film on the inside of their capsule when they come to a sudden stop.

CHAPTER 14

WHO ARE THEY?

IF THESE ALIENS and their craft are not actual physical beings, their bodies not made up of physical mass as we know of the physical properties of bodies, then they must be non-physical. As stated earlier, everything in our universe/cosmos has either mass or energy and is subject to all laws of physics. Even light, made up of photons, gives off or conducts energy.

Beings in the other dimension may be made up of entirely different mass properties. They could easily have the ability to appear or disappear whenever they wish. They could traverse between the dimensions instantly and without restrictions. They have demonstrated the ability of not being restricted by physical objects, like atmosphere, locked doors, walls, or anything else that are physical barriers to us. We cannot walk through a concrete wall or a steel door, but they apparently can. Not being bound by our physical laws, they can probably even change form, change themselves into or out of any form they desire, *making them very deceptive.* If they wanted to, they likely could even appear as *humans*. They could appear as aliens from some other planet, in some other galaxy, or anything else to facilitate the *deception* of making *us* believe that they are something different than what they really are.

Again, *are we humans being set up?* If this is the case, then *why* are they doing this?

How do I know so much about this other dimension? The truth is, these "other-dimensional" attributes have been communicated, telegraphed, and directly spoken to human beings and been recorded for several thousand years.

Here is what we already know about this "other dimension." This other dimension existed *before* our dimension/cosmos was created, with time being one aspect of our dimension. "Before the world/cosmos began" or existed and a reference to "before time began" comes from biblical books: John 17:5 and 1 Corinthians 2:7. The Bible is the *only* "Holy Book" that refers to another dimension, another existence before time (creation) began.

Eternity is a place where time does not exist. If someone is from eternity, they must be from this other dimension. Micah 5:2 speaks of a Ruler whose origins are from old, from eternity, who did—and will—come to Earth. He was and *is* a benefactor Being. He did—and will—come from this "other dimension." In this other dimension, there are benefactor beings and there are also *destructive* beings: beings who care about the human race and also beings who want to ultimately destroy—or remake—the human race. Isn't this the concern that Stephen Hawking has, that some of these UFOs may have ulterior and destructive motives concerning Earth and its human population? This concern is real. We need to know who we are dealing with and where they come from.

We learn from this ancient communication called "the Bible" that in this "other dimension" there are created beings, both good and deviant. They have bodies that are not subject to our physical laws. They have the ability to appear in any form they want, or to remain invisible (Judges 13:18–23; John 20:19–21). In this other dimension, there is a division of thought and struggles for power. The Supreme Being in this other dimension, who created all others within that dimension, has had a mutiny. The rebelling beings want to be "controllers" of that dimension. In the process, they want not only to depose the Supreme Being and His loyal followers, they also want to damage, destroy, and take control over—*if possible*—the *other* dimension, *our dimension*, created by the Supreme Being.

In particular, the deviant beings want to damage, destroy, or exercise control over humans on planet Earth, which was also created by the Supreme Being. Within these two "other-dimensional" groups, there is a hierarchy of command. A hierarchy ultimately demands a supreme, final authority figure within that group; an "alpha male," if you will. Most assuredly, the deviant beings *have* just such an authority figure—and he has a name and title.

The deviant beings have the ability to influence humans without their knowledge. They can implant ideas and knowledge into the consciousness of people or people groups, all while the affected individuals are thinking it was their own idea. Thought transference from the "other-dimensional" beings to humans in our dimension is instantaneous. With the right information on this "other dimension," we humans have the ability to distinguish between these two groups. Without the right information we have a great tendency to listen to the loudest voice, which comes from the deviant beings, all the while believing it is our own thoughts. *These deviant beings do not have our best interests at heart.*

This "thought transference" can be demonstrated in the book of Job, chapters one and two. In this narrative, the Supreme Being gives permission to the final-authority figure of the deviant group, to see if he can dissuade this human (Job) from being a follower of the Supreme Being. This high-ranking deviant being then influences a people group known as the Sabeans. The Sabeans (thinking it was their own idea) formed a raiding party, killed some of Job's workers, and carried off much of his livestock. Another people group called the Chaldeans came and killed some of Job's family and workers and carried off what was left of his possessions. These people groups did not have a clue that they were being influenced and prompted to do these things by a being from the "other dimension."

Ephesians 6:12 warns us about the influences of these "other-dimensional" beings, and tells us how to neutralize their promptings. The apostle Paul tells us that our struggle is not *ultimately* against humankind, but against "rulers" (in the Greek, high-ranking "spiritual" [read "other-dimensional"] beings) and "authorities" (in the Greek, the ability to control humans) in "heavenly" (other-dimensional) places. To paraphrase, our struggle is *not* against flesh and blood, but against other-dimensional beings that have the ability to control, prompt, and influence humans *from* this "other dimension."

What better way to deceive us concerning this Supreme Being, or even make us question His existence (as the world has often already been taught), than to someday make direct contact with planet Earth, claiming to be this Supreme Being (God), from another planet?!

There *is* evidence in Scripture that beings from the "other dimension" (referred to as "sons of God" in Genesis 6), *somehow modified the DNA of human offspring*, creating a race of men known as Nephilim ("fallen ones"). They were physically larger and seemingly much more intelligent than normal humans of that time. *They were hybrids! They had the ability to pass on their hybrid DNA.* This could have been *part* of the reason God destroyed the human race, except for eight people, with the flood during Noah's time. *The DNA of the human race had been corrupted.* These "sons of God" *were* the deviant, rebellious, and fallen beings from the other dimension, who had visited the Earth.

Today, every known species on the planet is to be considered for possible DNA transplants. Mixing and matching the genes of many different species, including humans as shown in the chapter on genetic engineering, *has already taken place*. It is a new form of breeding, of mating different species, to effect a specified outcome. Leviticus 19:19 tells us, "Keep My decrees. Do not mate different kinds of animals." And Leviticus 19:15–16 says "If a man has sexual relations with an animal" or "if a woman is inclined to have sexual relations with an animal" both people and animals are to be disposed of. *Is this not what the genetic labs are doing at the cellular level?* There is something inherently wrong with what scientists are doing in the field of genetic engineering.

Satan, the high-ranking spiritual being who led *and is leading* the rebellion against God—and who is the final authority figure of the fallen beings—has the ability to not only influence people, but to even invisibly *indwell* people. In John 13:2 we read, "The evening meal was being served, and the Devil had already prompted Judas Iscariot, son of Simon, to betray Jesus." This "prompting" occurs outside the body and mind to telepathically imprint thoughts and ideas into a human's consciousness, for the purpose of achieving the prompter's agenda. Verse 27 of the same chapter states: "As soon as Judas took the bread, Satan entered into him." Now you can see the difference between spiritually *prompting* a person and spiritually *entering into* a person.

There is coming a time in the not-too-distant future when a "special" human, a "super hero," *will* allow Satan to influence and then enter into, take possession of, and control of his body. He will be known to the world as "god" and "savior." He will be loved, respected, and a very likeable, charismatic person. Who knows, he might even show up in a UFO! He will attempt to prove who he is to the world by heretofore unseen feats of accomplishment that have previously been totally and completely impossible for humans to perform. He will accomplish many miraculous signs and wonders in his quest to convince people that *he* has all the answers ... that *he is the only hope the world has so as not to go into extinction.* People will follow him and do whatever he says for the sake of peace, safety, and sustenance for their families. Listen to what Scripture has to say about this "God wannabe" and ruler that *is* coming to the Earth:

> Don't let anyone deceive you in any way, for that day will not come until the rebellion occurs and the man of lawlessness is revealed, the man doomed to destruction. He will oppose and exalt himself over everything that is called God or is worshiped, so that he sets himself up in God's temple, proclaiming himself to be God.
> —2 Thessalonians 2:3–4

Remember earlier, when we noted that if one of these UFOs landed on the front lawn of the White House, it would cause mankind to rethink their views on religion and God? *We humans are being set up to believe a lie*, thus the probable reason for the proliferation of UFO movies, documentaries, and overall interest in the topic. Continuing with 2 Thessalonians 2:9–10: "The coming of the lawless one will be in accordance with the work of Satan displayed in all kinds of counterfeit miracles, signs and wonders, and in every sort of evil that deceives those who are perishing...."

People will be greatly impressed ... and deceived. Speaking of this "lawless one," Revelation 13:16–17 tells us, "He also forced everyone, small and great, rich and poor, free and slave, to receive a mark on their right hand or on their forehead, so that no one could buy or sell unless he had the mark...." The technology is available today to accomplish this kind of control over humanity. In the coming days, if you want to have a job and feed your family, you *will* "get with the program"—or die of hunger. You may even be killed if you don't "cooperate."

This "other dimensional" imposter knows that Jesus the Messiah, the Creator Spiritual Being, Who created all things, who Himself became a human, has promised to come back to the Earth to set up His kingdom for 1,000 years. In a last-ditch effort to beat Jesus to the punch, Satan will try to establish his kingdom on Earth first. At Armageddon, *Satan and the armies of the world are defeated by Jesus*. Jesus *will* set up *His* kingdom of *true* peace and provision, without conditions, for those who survive Armageddon.

However, *not many will survive the tribulation period* on the Earth, which will take place *before* Jesus establishes His kingdom of peace. Read and consider the words of Isaiah 24:1–6:

> See, the LORD is going to lay waste the earth and devastate it; He will ruin its face and scatter its inhabitants, it will be the same for priest as for people, for master as for servant, for mistress as for maid, for seller as for buyer, for borrower as for lender, for debtor as for creditor. The earth will be completely laid waste and totally plundered. The LORD has spoken His Word. The earth dries up and withers, the exalted of the earth languish. The earth is defiled by its people; they have disobeyed the laws, violated the statutes and broken the everlasting Covenant. Therefore a curse consumes the earth; its people must bear their guilt. Therefore earth's inhabitants are burned up, and very few are left.

This is the result of the seven-year tribulation period (described in the Book of Revelation) *and the final conflict at Armageddon*, before Jesus establishes His kingdom on Earth.

CHAPTER 15

THE PROPHECIES REVEAL ... IT *IS* GOING BY THE BOOK

TO IMPLY THAT the Antichrist would arrive by UFO may seem a little over the top. To imply that God would use of all the weapons of war mentioned previously could be a stretch. But one thing *is* certain: if man uses these weapons in a global war, or should God use some other, supernatural power to accomplish what's described in Isaiah 24, *the end result will be the same: human life will be greatly reduced.* (Also refer to Chapter 8 of this book, regarding nuclear radiation exposure, as well as the prophecy recorded in Zechariah Chapter 14.)

If there is no God, mankind will use these weapons and the end result will be the same. Regardless of whether God uses *His* means or mankind uses *its* means, still the words of Jesus remain ... "if those days had not been cut short no flesh (humans) would survive." If God does not intervene, the outlook for humankind is probably extinction. How long will it be before these things happen? No human knows.

But I think we can safely say less than 50 years are left, with the possibility of much less time, all converging things considered. We all had better hope there *is* a God and that He *has* indeed communicated to mankind the outcome of the world ... *and the way to escape that outcome.*

If it goes the way Scripture say it will go, then the masses of the Earth are in for increasingly great deception and a great reduction of people. It is possible that the UFO phenomenon will play a part ... maybe even a big part ... in this great deception. No one at this point can say. While I am not trying to imply it will be a UFO, the deception that is coming will be on a scale equivalent to a UFO landing on the White House lawn. It will be just as overwhelmingly believable. *Whatever this great deception is, the point of it is to lead people away from believing in and having faith in the God of the Bible.*

Concerning the coming man of lawlessness, 2 Thessalonians 2:4–10 says:

> He opposes and exalts himself over everything that is called God or is worshiped, and even sets himself up in God's Temple, proclaiming himself to be God. Don't you remember when I was with you I used to tell you these things? And now you know what is holding him back, so that he may be revealed at the proper time. For the secret power of lawlessness is already at work; but the One who now holds it back will continue to do so until He is taken out of the way. And then the lawless one will be revealed, whom the Lord Jesus will overthrow with the breath of His mouth and destroy by the splendor of His coming. The coming of the lawless one will be in accordance with

the work of Satan displayed in all kinds of counterfeit miracles, signs and wonders, and in every sort of evil that deceives those who are perishing. They perish because they refused to love the truth and so be saved.

People will believe the lie. They will be ripe for the deception. What would it take to make *you* believe this guy is the long-awaited "God"? What would he have to do to convince *you* to give *him* all authority over Earth? This deception will be so believable that it will deceive "even the very elect, if that were possible" (Matthew 24:24). The closer we get to the end times and the second coming of Jesus, the greater the crescendo of deception.

There have been many Scripture verses used so far to show the parallel of current human events and the writings of Scripture. These parallels that have been shown so far could *not* have been the product of "guess work" written 2,000 years ago, any more than *you* could exactly predict what will happen 2,000 years from now. And yet, they are all coming about, with the exact information describing these current days. The days yet to come are also described in Scripture. We can see firsthand, and *know* in our hearts, that these many predicted/prophesied things *have* happened, and therefore are very likely to *continue* to happen, just as it has been recorded.

As recorded in the book of Deuteronomy, Moses told the Israelites just before they were to enter the "Promised Land" of Canaan (having escaped from Egyptian bondage), that if they didn't follow the laws and decrees of God Almighty, that God Himself would see to it that the Israelites would be removed from the Promised Land. God said the Land would lie desolate for centuries, but in the "end of days" He would gather the Israelites from the four corners of the Earth, from the most distant lands, back to their homeland of Israel. Deuteronomy 30:1–5:

> When all these blessings and curses I have set before you come upon you and you take them to heart wherever the LORD your God disperses you among the nations, and when you and your children return to the LORD your God and obey Him with all your heart and with all your soul according to everything I command you today, then the LORD your God will restore your fortunes and have compassion on you and gather you again from all the nations where He scattered you. Even if you have been banished to the most distant land under the heavens, from there the LORD your God will gather you and bring you back. He will bring you to the land that belonged to your fathers, and you will take possession of it. He will make you more prosperous and numerous than your fathers.

The Jewish people were kicked out of the land by the Roman Empire in 70 A.D. and Israel was not re-established as a Jewish homeland until 1948. Jews have been going back to Israel from all over the world ever since.

When the Roman army surrounded Jerusalem in 67 A.D. and laid siege to it, there was estimated to be up to 3,000,000 people inside the city. Only 97,000 survived and were deported all over the Roman Empire. Speaking of Jesus, Luke 19:41–44 reads:

THE PROPHECIES REVEAL ... IT *IS*
GOING BY THE BOOK

> As He approached Jerusalem and saw the city, He wept over it and said, "If you, even you, had only known on this day what would bring you peace, but now it is hidden from your eyes. The days will come upon you when your enemies will build an embankment against you and encircle you and hem you in on every side. They will dash you to the ground, you and the children within your walls. They will not leave one stone on another, *because you did not recognize the time of God's coming to you*" (emphasis added).

Jesus said this just before His crucifixion, and it was approximately 40 years before His prophecy was fulfilled. You can read this same prophecy in Deuteronomy 28:49–63. It is a graphic prophecy of what would (and did) happen in Jerusalem. The Jewish historian Flavius Josephus, an eyewitness, recorded historical information about these things.

In history, whenever a people group or a nation have been displaced from their homeland like the Israelites have been, in every case, over time, those people groups were assimilated into the populations where they were relocated to. They would lose their language, their culture, and their religion after a few generations. It is historical fact that the Israelites are the *only* people group that have been displaced for nearly 2,000 years who kept their culture, kept their religion, kept their language, and kept their national identity, just as God and the Bible said they would, 3,200 years ago. No other so-called holy book demonstrates these fulfilled, historical events as does the Bible. There is no so-called "eyewitness" to the historical re-establishment of the State of Israel who inserted this prophecy into the Bible to make it come true, as the Bible's critics claim about other fulfilled, historical prophecies of centuries past. God proclaimed that this would be an "end time event." *This is how we know we are now living in what must be the end of the "end times."*

Never in the history of civilization has a people group like the Israelites been so hated from one generation to the next, from one century to the next, from one millennium to the next, like the people of Israel have. Never has another people group so many times been defeated, plundered, killed, forced into slavery, and scattered to the four winds by conquering armies, yet still keep their national identity like the Israelites.

They were hated and enslaved by the Egyptians. They were hated and defeated by the Assyrians in 722 B.C. Those who survived were taken back to Assyria as slaves and concubines. The job was completed with the destruction of the city of Jerusalem and its temple by the Babylonian Empire. Captives were taken back and placed all over the Babylonian Empire. The Persian Empire then conquered the Babylonian Empire, which allowed the Israelite captives to go back to their homeland and rebuild their temple. They were, however, still part of the Persian Empire and subject to Persian rule. Then came Alexander the Great of Greece. He defeated the Persian Empire, marched into Jerusalem and on into Egypt, where he built a new city named after him, Alexandria. Israel was then under Greek rule and laws. It was a time of great persecution and many Jews were scattered all over the Greek Empire.

THE PROPHECIES REVEAL ... IT *IS* GOING BY THE BOOK

The Roman Empire dispossessed the Greek Empire in 63 B.C. and became the hated rulers of Israel. Persecution was great in the land of Israel. Caesar was god. Anybody claiming to be God was a threat to Caesar, and that was punishable by death. It was during this time that Jesus was put to death on a cross, while affirming that He was the Son of the one true God.

In the Old Testament book of Daniel, every one of these empires was named in succession, as was the death of the Anointed One, the Messiah. All were prophesied by Daniel, beginning around 550 B.C. The angel Gabriel told Daniel that one of these events would culminate in the death of the Messiah. Daniel lived long enough to see the first event: the Persian Empire taking over the Babylonian Empire, without firing a shot. This was also prophesied in the book of Isaiah. Moses prophesied in Deuteronomy 28, Daniel prophesied in Daniel 9 and Jesus prophesied in Luke 19 that the Romans would once again destroy the city and the temple. This happened in 70 A.D. and the Jews were scattered to the four corners of the Roman Empire. Some escaped to India and others to Asian countries.

This should have been the end of the Jews' story. *Obviously it wasn't.* These things are recounted here to show the amazing accuracy of biblical prophecy and the absolute *impossibility* of a people group being able to maintain their religious, linguistic, and national homeland aspirations, *against all these odds*, without divine intervention. The historical story of the Jewish people, encompassing past, present and future, *is not over.*

Beginning about the year 1000 A.D. and continuing until the recent past of the 20th century, the Jews began to be severely persecuted and murdered worldwide. The new Muslim religion, Islam, founded some 300-plus years prior to 1000 A.D., was killing Jews by the tens of thousands and enslaving those that remained. The Jews found themselves being forced to migrate all over Russia, Poland, Germany, France, Spain, Italy, and England. If some Jews in any town or area were killed and the remainder of them were then chased out of that area, they would go to another town and re-establish with what remained. They were repeatedly and falsely accused of being Christ-killers, baby-killers, creators and spreaders of the Black Plague, of plotting to take over the world, and accused of poisoning wells when people died mysteriously.

Not only were they murdered and run out of towns and cities, they were also murdered and run out of countries by the tens of thousands each time. Their properties were confiscated or burned. They were killed during the Spanish Inquisition if they didn't convert to Christianity. There were six million Jews killed by Hitler in World War II. *And it was all prophesied in the Bible.*

Following is an excerpt of that prophecy. This prophecy was written around 1250 B.C. by Moses. It covers the almost 2,000-year dispersion from the Jews' homeland of Israel, chased out by the Roman army. Deuteronomy 28:63(b)–67 reads:

> You will be uprooted from the land you are entering to possess. Then the LORD will scatter you among all nations, from one end of the earth to the other. There

you will worship other gods—gods of wood and stone, which neither you nor your fathers have known. Among those nations you will find no repose, no resting place for the sole of your foot. There the LORD will give you an anxious mind, eyes weary with longing and a despairing heart. You will live in constant suspense, filled with dread both night and day, never sure of your life. In the morning you will say, "If only it were evening!" and in the evening, "If only it were morning!"—because of the terror that will fill your hearts and the sights that your eyes will see.

Put yourself in their shoes ... would you not live in constant suspense, not being sure if you or any of your family were going to live from one day to the next? Would you not come to a point, like modern Jews, and say "No more!" Would you not say, "We will fight before all of that horror happens again?"

After all, God *did* promise to restore the Jews to their homeland of Israel at the end of time and *we are eyewitnesses of that event.* On May 14, 1948, the modern State of Israel was re-born ... and immediately had to fight a war against the Arabs (a war threatened and started by the Arabs), for its very existence. That has happened repeatedly in the ensuing years. And now, the Arab/Islamic nations of the Middle East—and soon *all* the nations of the world—are plotting and *will* come together in an attempt to accomplish what the world has failed to accomplish throughout the centuries ... to exterminate once and for all the Jewish people and occupy the entire Land of Israel; all of which is also prophesied in the Bible. *It is going by the Book!*

The Islamic nations listed in Ezekiel 38 and Psalm 83 have vowed to wipe Israel off the face off the Earth. Listen to Psalm 83 and tell me you haven't heard or read this threat in the media. If you haven't, you have not been paying attention!

> O God, do not keep silent; be not quiet, O God, be not still. See how Your enemies are astir, how Your foes rear their heads. *With cunning they conspire against Your people*; they plot against those You cherish. "Come," they say, *"let us destroy them as a nation, that the name of Israel be remembered no more." With one mind they plot together; they form an alliance against You.*
> —Psalm 83:1–4, emphasis added

Never in history has this coalition of all Middle Eastern nations come together with one heart and one mind to destroy the nation of Israel ... *until now*. It has been said that if the Islamic nations of the Middle East would lay down their arms there would be peace, but if the Israelis lay down their arms, there would be no Israel.

One thing is for sure: there is much, much more to this conflict than what has been *and is being* reported in the major media, worldwide. All of the biblical prophecies forecasting that *all* of the Islamic Middle Eastern nations (and possibly also Russia) coming against Israel, in a war that will occur *before* the war of Armageddon (which will include *all* nations; reference Joel 3), would be worthless, *if Israel was not in their homeland.*

Israel *has* to be in the Land *before* the prophetic promises can be fulfilled, and against all impossible odds, they are back in the Land. In 1948 Israel became a nation, fulfilling Moses' prophecy recorded in Deuteronomy 30 and also other biblical prophecies.

This time the Jewish people *will* fight, and fight the whole world if necessary. Even though they will be overwhelmingly outnumbered and overwhelmingly outgunned, *God says He will intervene* to secure their survival.

Woe to those who fight against the Jews, because they will be fighting against God!

What is really going on here? How can this be? Why is the whole world focused so intently on a tiny country of seven million people? Why has this always been true for centuries and even millennia? This is not reasonable or logical. Israel is a prosperous little country reborn out of the ashes. Israel has the ability to help the nations surrounding it economically, *if* they would let it.

However, the issue *isn't* economics, it's *religious*. The two religions represented here, Judaism and Islam, are diametrically opposed. This is very similar to the opposition that's going on in the other dimension, in heaven.

Mark Twain, a.k.a. Samuel Clemens, visited the Land of Israel in the 1800s and remarked that he didn't understand why anybody would want to live there, because the place was so desolate. God promised a restoration of the Land when He began restoring His people to the Land. Israel is anything but desolate now! Its restoration began in earnest in 1948.

So why do we see such intense worldwide hatred of the Jews and Israel? There really can be only one logical answer. It is the ability of Satan—the high-ranking, created spiritual being from the other dimension, who is rebelling against God Himself and trying to destroy *everything* that is important to God ... including His chosen people, the Jews—to influence any number of people groups at any given time to perform *his* dirty work, often without the knowledge of the people he is working through!

CHAPTER 16

THE CONCLUSION

IT HAS BEEN demonstrated in this book that in the last fifty years there has been an increase in large earthquakes and especially so in the last ten years. How intense will these earthquakes become in the future? This is what is revealed in Revelation 16:18–19: "Then there came flashes of lightning, rumblings, peals of thunder and a severe earthquake. No earthquake like it has ever occurred since man has been on the earth, so tremendous was the quake. The great city [Jerusalem] split into three parts, and the cities of the nations collapsed." Not a pretty picture.

It has been demonstrated that severe weather events around the world are on a sharp increase, which the Bible also predicts. Severe weather-related death and destruction, as well as death and destruction from all manner of natural disasters, have increased sharply worldwide, especially in the last ten years. This includes blizzards, wildfires, floods, windstorms, tornadoes, hurricanes, tsunamis, and droughts.

One only has to watch the evening news to know this is true. More telling than the evening news are the millions of people worldwide each year who have suffered through these events.

Volcanic activity has increased. There is a biblical prophecy (Revelation 8:8–9) concerning an asteroid the size of a mountain striking the Earth, though in the first century when this was written they didn't even know what an asteroid was. They only knew that it was huge, all aglow with fire and coming from the heavens. They didn't know what nuclear radiation was, yet 2,500 years ago the Prophet Zechariah described its symptoms perfectly (Zechariah 14:12).

These are *all the beginnings* of "birth pangs" spoken of by Jesus, reserved for the end times. This kind of foreknowledge, recorded many, many centuries ago, could *only* come from an information and knowledge source outside of our dimension, where time is non-existent.

There are great fears among the medical community that a viral pandemic, such as what occurred in 1918 that killed millions of people worldwide, will occur again. When we consider all of the deadly pathogens stockpiled by the nations around the world, and the possibilities of their use in wartime or by terrorists, it is really not a matter of *if* but *when* these virulent pathogens will be released. These virulent pathogens can be either naturally-occurring or man-made or both. The outcome is the same: mass death.

Jesus said in Luke 21:11, "There will be great earthquakes, famines and pestilences in various places, and fearful events and great signs from heaven." Notice that earthquakes, famines, and pestilences are all *plural*. There will be *many* of each kind, in *many* locations. The dictionary defines pestilence as a "contagious or infectious epidemic disease that is virulent and devastating; especially the Bubonic Plague." (Merriam-Webster Collegiate Dictionary, 10th Edition). This disease alone wiped out one-third of Europe's population before it ran its course. These two, earthquakes and pestilences, can by themselves cause major famines around the world. *These events are to take place during the end times, just before the return of Jesus.*

Except for the possibility of deadly pathogens being released by man, the previous paragraphs are only highlighting the natural disasters that are already occurring in ever-increasing numbers. This book has also highlighted the possibility of man-made disasters that are likely to occur. Without going into detail again, these man-made disasters include cyber warfare, nuclear proliferation, dwindling natural resources, and population explosion. These are all *converging* to an outcome that will be devastating to the Earth's population. Mankind will not be able to keep these four man-made components from reaching their natural conclusion.

Let us consider again the issue of Israel and the Middle East, where all the pieces of the biblical puzzle of nations seem to be coming together. For so many centuries, people, including learned Christians, were not able to believe for various reasons that this prophecy of Israel repopulating its ancient homeland could come to pass. Yet it has indeed and here we are *today*, with Israel in the Land!

If what Jesus says was and is true, then what did *He* have to say about the days leading up to the seven-year great tribulation period? In Luke 17, beginning with verse 26, He said it will be just like it was in the days of Noah and Lot: "People were eating, drinking, marrying and being given in marriage ... buying and selling, planting and building ..." up to the day Noah entered the ark and Lot left Sodom. That isn't anything unusual-sounding in the day-to-day life of the people of that day.

However, if we go back to Genesis and see what it was *really* like in the days of Noah and Lot, above and beyond what Jesus spoke of in Luke, a more complete picture emerges. Genesis 6:1–2, 4–5 and 11–12:

> When men began to increase in number on the earth and daughters were born to them, the sons of God saw that the daughters of men were beautiful and they married any of them they chose.... The Nephilim were on the earth in those days—and also afterward—when the sons of God went to the daughters of men and had children by them. They were the heroes of old, men of renown. The LORD saw how great man's wickedness on the earth had become and that every inclination of the thoughts of his heart was only evil all the time.... Now the earth was corrupt in God's sight and was full of violence. God saw how corrupt the earth had become, for all the people on earth had corrupted their ways.

Apparently the "sons of God," who many believe were the fallen, rebellious beings from the other dimension, somehow had the ability to corrupt the DNA of humans. The corruption of this human DNA led to physical giants in the land, with higher-than-normal mental skills. The embryonic changes made to the reproductive organs of the women became hereditary and were passed on from generation to generation. With the Earth being full of corruption and violence, with the thoughts of men being constantly evil, the sexual perversion displayed in Genesis 19 toward Lot, his family, and the two visitors that came to see him, all of this *may* have been the result of the corrupted DNA and was more than God would tolerate.

Out of all of the population of Earth at that time, only Noah and his family were considered to be uncorrupted. The point here is that Jesus said, "this is how it will be" leading up to His return, at the end of the tribulation period. That corruption, violence, sexual perversion, and the corruption of the DNA of all species *will* be happening. It will all be considered just as "normal" as buying and selling, marrying and being given in marriage, eating and drinking. Isn't all this just what we are experiencing today?

Violence, corruption, and sexual promiscuity are all rampant in the world's society today. Morality and honesty were the hallmarks of moral integrity and were the norm in past generations. But in the last 50 years violence, corruption, and sexual promiscuity have become the norm. It has become so common—and we have become so desensitized to it—that it no longer seems out of place.

What *does* seem "out of place" to most in our society are those who espouse biblical authority, integrity, and morality. The "Bible thumpers" are said to be really out of touch with reality … *or are they?*

If we are really honest about it, there is rampant corruption in government, corruption in the banking system, corruption in business, law enforcement, education, the church, and industry. I once had a CEO say to me that he would rather steal the technology than pay to have it developed! He is certainly not alone in his view. This "pandemic" of corruption, violence, and sexual perversion is *worldwide*.

Drug violence is also an epidemic worldwide. Not only with drug cartels, but every-day junkies on the street are killing each other. There are several states in the United States where you are more likely to die from a gunshot than from an automobile accident.

Considering the standard the Bible holds up on sexual morality, sexual perversion is off the charts. After all, "if there is no God, then there is no right or wrong"—therefore, there is no *moral* authority to answer to. Ultimately, this belief system will lead to civil disobedience, or worse. History has proven this true. That is where we are headed.

Another indication of end times is Christian persecution and martyrdom, along with anti-Semitism (hatred of Jews). The number of Christians who have been killed for their faith, which was upwards of 100,000 in 2011, is the highest since the first century. And the number of Jews who are killed simply because they are Jews also continues to climb.

In this conclusion we have yet to mention global economic turmoil, which *will* lead to a one world economy. We are almost there now! We also have yet to mention the commonplace, widespread disregard for laws and for those in authority, the breakdown of families and family values, and the rejection of biblical authority, *even within the Church*.

THE CONCLUSION • 153

All of the different, specific things we see taking place in the world today have been prophesied in Scripture. How many times does the Bible have to be right in its prophetic pronouncements before *we* figure out that it is not only right, but it has been 100% right all along?

If you choose to *not* believe the Bible and instead prefer to trust mankind to get us out of the *converging catastrophes* through technology and ingenuity, here is what you can expect. Mankind will continue to super-enhance humans through genetic engineering and computer/human interfacing, probably in less than 50 years. That is the Transhumanists' estimate, not mine. That is, they will do so if they can avoid the wars.

One thing Transhumanists and genetic engineers understand, is that the longer humanity lives and the healthier we become, so that death is almost unheard of, the fewer people this planet can support. Ideally, researchers who study these demographics suggest that *no more than 500 million people* be permitted to live on the planet at any one time, due to limited resources! *This implies that the six and one-half billion excess people now residing on this planet will have to find some other place to live, or be eliminated.* I might add that this elimination would have to be done rather quickly. How do you suppose those in control could mastermind such an event? Here is one way: they could develop a lethal virus, then develop a vaccine for that virus, secretly give it to the elite 500 million ... and then release the virus on the unsuspecting world. After all, the medical elite have been telling us for some time that a lethal, viral pandemic could strike the Earth with catastrophic effects. Who would expect or suspect such a conspiracy, until it is too late?

How many guesses do you want to take regarding which group you will be in? The people in authority, the highly-educated, the rich and powerful, those with "connections" will almost certainly make up the 500 million. Where will *you* be in this preferential list?

If computers become self-aware and self-assembling, and decide these 500 million people are a threat to *them*, what do you think would be the outcome for the 500 million? Can't happen? What, do you think the UFOs are going to come and save them? I guess anything is possible, but will it be too late for you and your family? If you are not among the "rich and famous," you will not escape.

Whether I am right or wrong, the prognosis for you is bleak. If this futurist scenario in any form is what the Earth has to look forward to, you and your family are not going to make it. *If the scenario described in the Bible is what the Earth has to look forward to and if you and your family are not practicing, believing Christians, you are not going to make it. There is no escape for you under the Transhumanist agenda.*

Here is the plain truth, if you are willing to hear it and then choose to accept it: *There is an escape for you under the biblical scenario, no matter who you are, if you believe and trust in Jesus.* Jesus said "I am the Way, the Truth and the Life; no one comes to the Father except through Me" (John 14:6). This is either the statement of a very arrogant man, or it is the truth! Jesus also said in Luke 21:35–36: "For it will come upon all those who live on the face of the *whole* earth. Be always on the watch and pray that you may be able to *escape* all that is about to happen and that you may be able to stand before the Son of Man" (emphasis added).

Should you decide to become a follower of Jesus Christ, here is some of what you can expect as the future unfolds. If you live in any of the Muslim countries, you can expect to be severely persecuted and/or killed for your belief. It is happening even as this is being written. If you live in any of the so-called developed countries, you can expect to be ridiculed, marginalized, and prevented from speaking biblical truths in public venues of all kinds. You will be excluded from some jobs, especially teaching jobs. This will all come from both the public and private sectors. You will be told that you have to be tolerant of other people's views, but your views will not be tolerated. Your civil and constitutional rights will be eroded to the point where you will be arrested for speaking biblical truths ("hate speech"), or conducting or attending biblical meetings and gatherings.

You will possibly have to suffer, with the rest of mankind, a large but somewhat-limited war, centered around Israel and the Middle East, *before* the seven-year tribulation period begins. Before that tribulation period begins, Jesus will come and gather up His Church, made up of true believers like you. He will take you to heaven for the duration of the great tribulation, which culminates with all nations of the Earth coming against Israel. Israel would not survive if Jesus did not come back to defend it and to defeat Satan.

After that time, Jesus will renew and restore the Earth to pristine condition. The entire creation will be set free from the bondage of sin and death (Matthew 19:28; Romans 8:18–22). You and other believers in Jesus, along with the few survivors of the tribulation period, will enjoy a renewed and restored Earth for 1,000 years (Rev. 20:1–10). After that period of time, God will create a *new* heaven and earth.

God provided a way for Noah to escape the wrath that was about to come upon the earth. He and his family were sealed in the boat (Ark) seven days before it started to rain.

God *will again* provide a way to escape all that is about to happen ... *for those who believe and trust in His Son.* This is known as the "rapture." This is how it will happen according to 1 Thessalonians 4:13–18:

> Brothers, we do not want you to be ignorant about those who fall asleep, or to grieve like the rest of men, who have no hope. We believe that Jesus died and rose again and so we believe that God will bring with Jesus those who have fallen asleep in Him. According to the Lord's own word, we tell you that we who are still alive, who are left till the coming of the Lord, will certainly not precede those who have fallen asleep. For the Lord Himself will come down from heaven with a loud command, with the voice of the archangel and the trumpet call of God, and the dead in Christ will rise first. After that, we who are still alive and are left will be caught up with them in the clouds to meet the Lord in the air. And so we will be with the Lord forever. Therefore encourage each other with these words.

If the Bible is wrong and my personal trust in Jesus is misplaced, then my fate will be the same as yours, if you do not personally believe in Jesus.

THE CONCLUSION • 157

However, if the Bible *is* right (as I believe it has been demonstrated in this book to be), *then what will your fate be*? Are you *really* willing to gamble your life, and possibly the lives of your family, believing that there is no God and that mankind will solve its own problems?

It is time to decide which you will believe.

Good luck ... or, God's blessing to you ...
it depends on your choice.
I sincerely pray you will choose the right way.

www.ingramcontent.com/pod-product-compliance
Lightning Source LLC
Chambersburg PA
CBHW030324080526
44584CB00012B/702